University of South Wales

2072420

An introduction to

Engineering
Mechanics

An introduction to
Engineering
Mechanics

DAVID REID

BSc, MSc, PhD, CEng, CMath, MICE, FIMA
Senior Lecturer, Napier University, Edinburgh

palgrave

© David Reid 2001

All rights reserved. No reproduction, copy or transmission of
this publication may be made without written permission.

No paragraph of this publication may be reproduced, copied or
transmitted save with written permission or in accordance with
the provisions of the Copyright, Designs and Patents Act 1988,
or under the terms of any licence permitting limited copying
issued by the Copyright Licensing Agency, 90 Tottenham Court
Road, London W1P 0LP.

Any person who does any unauthorised act in relation to this
publication may be liable to criminal prosecution and civil
claims for damages.

The author has asserted his right to be identified
as the author of this work in accordance with the
Copyright, Designs and Patents Act 1988.

First published 2001 by
PALGRAVE
Houndmills, Basingstoke, Hampshire RG21 6XS and
175 Fifth Avenue, New York, N.Y. 10010
Companies and representatives throughout the world

PALGRAVE is the new global academic imprint of
St. Martin's Press LLC Scholarly and Reference Division and
Palgrave Publishers Ltd (formerly Macmillan Press Ltd).

ISBN 0–333–93079–7 (paperback)
ISBN 0–333–94921–8 (hardcover)

This book is printed on paper suitable for recycling and
made from fully managed and sustained forest sources.

A catalogue record for this book is available
from the British Library.

Library of Congress Cataloging-in-Publication Data

Reid, David.
   An introduction to engineering mechanics/David Reid.
      p.   cm.
   Includes bibliographical references and index.
   ISBN 0–333–94921–8
   1. Mechanics, Applied.   I. Title: Engineering mechanics.   II. Title.

TA350 .R395 2000
620.1 – dc21   00-50118

10   9   8   7   6   5   4   3   2   1
10   09   08   07   06   05   04   03   02   01

Printed in Great Britain by
Anthony Rowe Ltd, Chippenham, Wilts

Contents

Preface

Existing engineering mechanics text-books for undergraduate students tend to treat dynamics and statics separately. In doing so they miss the opportunity to develop the student's ability to see the subject in its entirety. In particular, students tend to regard statics as a subject unrelated to the fundamentals of dynamics. Not only does this lead to additional effort being needed in assimilating the principles underlying the two areas, but it also denies the student the opportunity to develop a deeper understanding of the concepts involved. In this text the two areas are fully integrated and statics is seen to be a special subset of dynamics where Newton's equations of motion are set equal to zero as a result of equilibrium considerations.

The book is aimed at students beginning an undergraduate course in any of the branches of engineering where an understanding of engineering mechanics is an essential element. It is assumed that these students will have completed Mathematics or Physics at A level or Scottish Higher. The text covers the range of material expected in the first semester of year 1 and indeed is drawn from the author's experience of presenting such material on the undergraduate courses in Civil Engineering at Napier University.

Throughout the book, the importance of starting any problem by drawing a free-body diagram is emphasised. If this fundamental principle is grasped then the student will gain the necessary confidence required to analyse any engineering mechanics problem, rather than merely a facility to reproduce solutions to problems which are similar to those already encountered.

A particularly strong feature of the book is the number and range of fully worked examples. A full description of each step in the solution is given in each case, with a review of the major points given at the end of each chapter or section. Theory is kept to a minimum but an explanation and review of the equations and concepts used in the examples are given where these are essential for proper understanding.

Chapter 1 deals with the basic concepts, much of which will be revision for the majority of students. Single- and two-dimensional kinematics is covered in Chapter 2. Chapters 3 and 4 introduce Newton's laws, starting with simple translational examples and progressively developing more involved and complex systems. Statics, in the form of equilibrium and moments, is shown to be a logical consequence of the laws in specific cases. Chapters 5 and 6 are devoted to the important topics of work, energy, impulse and momentum. Chapter 7 deals with the formulation of beam equations and how these can be used to determine maximum values and points of contraflexure on simple statically determinate systems. The elastic curve method is also introduced, leading to calculations involving beam slope and deflection. The final chapter is an introduction to vibration theory. The examples are restricted to simple one degree of freedom, undamped systems subjected to sinusoidal forcing vibrations.

D.B. Reid

List of Symbols

a	$m\,s^{-2}$	translational acceleration
a	m	breadth
A	m^2	area
b	m	depth
C		constant of integration
C	m	response amplitude
e	m	extension
e		coefficient of restitution
E	$N\,m^{-2}$	Young's modulus
E	J	mechanical energy
f	Hz	natural frequency of vibration
F_0	N	driving force amplitude
G	$N\,s$	linear momentum
h	m	axis shift distance
h	m	relative height
I	$kg\,m^2$	second moment of mass
I_A	m^4	second moment of area
J	$N\,s$	impulse
k	$N\,m^{-1}$	stiffness
k	m	radius of gyration
K	J	kinetic energy
l	m	length
m	kg	mass
M	$N\,m$	moment
r	m	radius
R	N	reaction force
R	m	radius of curvature
s	m	distance
S	N	friction force
S	N	shear force
S	J	strain energy
t	s	time
T	$N\,m$	torque
u	$m\,s^{-1}$	initial translational velocity
v	$m\,s^{-1}$	final translational velocity
V	m^3	volume

w	Nm^{-1}	uniformly distributed load
W	N	point load
W	N	weight
x	m	position
X_0	m	forcing displacement
Y	m	deflection
α	$rad\,s^{-2}$	angular acceleration
δL	m	change in length
Δ	J	mechanical work
ε		strain
μ		coefficient of friction
θ	rad	angular displacement
θ	rad	slope
ρ	$kg\,m^{-3}$	density
σ	Nm^{-2}	direct stress
τ	s	period
ω	$rad\,s^{-1}$	natural frequency
ω_1	$rad\,s^{-1}$	initial angular velocity
ω_2	$rad\,s^{-1}$	final angular velocity
Ω	$rad\,s^{-1}$	forcing frequency

CHAPTER 1

Initial Concepts

Traditionally, engineering mechanics can be considered to consist of two fundamental areas: dynamics and statics. *Dynamics* is the study and analysis of bodies in motion and the forces that produce these motions. *Statics* is the study of the forces on bodies in equilibrium, i.e. where there is no movement. In fact, as will be seen, statics is really just a special case within the overall topic of dynamics and that is how it is treated in this text. The term *kinematics* refers to the analysis of the motion of bodies without concern for the forces that produce these motions.

In this first chapter we will review the following concepts which form the foundation to a study of engineering mechanics:

- *Scalar, vector, mass, force, weight, volume, density, angle*
- *Force components*
- *Moment*
- *Torque*
- *Differentiation*
- *Integration*
- *Centre of mass*
- *Second moment of mass*
- *Radius of gyration*
- *Centroidal axes*
- *Second moment of area*
- *Parallel axis theorem*
- *Direct stress*
- *Direct strain*
- *Elasticity*
- *Young's modulus*

1.1　Basics

1.1.1　Scalar/vector

It is essential from the outset to distinguish between quantities which have magnitude only – *scalars*, and those which have magnitude associated with a direction – *vectors*.

1.1.2　Mass

Mass (m; kg) is the quantity of matter in an object. Mass has magnitude but no direction. It is a scalar.

1.1.3　Force

Force (F; N) is a measurable and determinable influence inclining a body to motion. In order to cause a body to move or cause a change in its movement, a force must act upon that body. Force is a vector. It has magnitude and direction.

1.1.4　Weight

Weight (W; N) is a force and is equal to an object's mass multiplied by the acceleration due to gravity (see also Newton's laws in Chapter 3). On Earth, weight (N) is equal to mass (kg) multiplied by 9.81 and acts vertically downwards. Weight is therefore a vector quantity.

1.1.5　Volume

Volume (V; m^{-3}) is the space occupied by an object. Volume is a scalar.

1.1.6　Density

Density (ρ; $kg\,m^{-3}$) is a measure of the quantity of matter in each unit of volume and is a scalar.

$$m = \rho V$$

(1.1)

EXAMPLE 1.1

Determine the mass and weight of a concrete column $1.5\,m \times 1.25\,m \times 8\,m$ high. The density of the concrete is $2400\,kg\,m^{-3}$.

Solution

$$V = 1.5 \times 1.25 \times 8 = 15\,\text{m}^3$$
$$m = \rho V = 2400 \times 15 = 36\,000\,\text{kg}$$
$$W = mg = 36\,000 \times 9.81 = 353\,160\,\text{N} = 353.16\,\text{kN}$$

1.1.7 Angles

Angles can be measured in degrees (°) or radians (rad). The relationship between these alternative forms can be deduced from the diagram shown at right.

A *radian* is the angle subtended by a circular arc whose curved length is equal to the circle radius, r. The circumference of the circle is equal to 2π multiplied by r and there are 360° in one complete revolution of the circle. Hence

$$2\pi \ (\text{radians}) \equiv 360\,° \equiv 1\ \text{revolution} \tag{1.2}$$

1.2 ▌ Force components

In many of the problems that arise in engineering mechanics it is convenient to express the effects of a single force as the sum of two parts or components. Since forces are vectors they have magnitude and direction. Forces can be represented graphically by a line drawn in the direction of the force and with a length, to a convenient scale, representing its magnitude.

From the basic principles of vector addition, force F, shown at right, is the sum of F_1 and F_2.

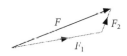

F can be replaced by any two (or more) components F_1 and F_2 provided that when F_1 and F_2 are set down, head to tail, as shown, they complete a triangle closing on F.

Generally, the best choice of components for F are those which form a right-angled triangle on F.

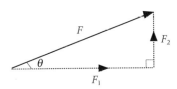

The following trigonometric relationships are then evident:

$$\begin{aligned} F_1 &= F\cos\theta \\ F_2 &= F\sin\theta \\ \theta &= \tan^{-1}\left(\frac{F_2}{F_1}\right) \end{aligned} \tag{1.3}$$

And from Pythagoras:

$$F = \sqrt{F_1^2 + F_2^2}$$

(1.4)

1.3 Moment

The *moment* (M; Nm) of a force about a point, A, is defined as the magnitude of the force multiplied by the perpendicular distance between the line of action of the force and the point, A. It is usual to consider clockwise rotations as positive and anticlockwise rotations therefore as negative. Moment is a vector.

In each of the following examples, determine the total moment at the point A due to the applied forces.

EXAMPLE 1.2

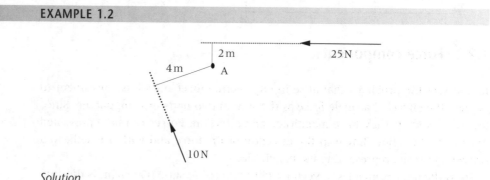

Solution

Take each force in turn and multiply the force by the perpendicular distance from its line of action to A. Treat clockwise rotations as positive and anticlockwise rotations as negative.

$$M_A = (10 \times 4) - (25 \times 2) = -10 \, \text{Nm}$$

The total moment at A is 10 Nm anticlockwise.

EXAMPLE 1.3

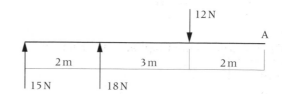

Solution

$$M_A = (15 \times 7) + (18 \times 5) - (12 \times 2) = 171 \, \text{Nm}$$

The total moment at A is therefore 171 Nm clockwise.

EXAMPLE 1.4

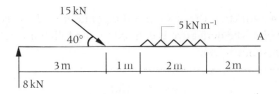

Solution

The 15 kN inclined load is split into its horizontal and vertical components in order to easily determine the perpendicular distance to A. Only the vertical component creates a moment about A. Consider the uniformly distributed 5 kN m^{-1} load as an equivalent point load acting at its centre.

$$M_A = (8 \times 8) - (15\sin 40°)(5) - (5 \times 2)(3)$$
$$= -14.209 \text{ kN m}$$

Hence the total moment at A is 14.209 kN m anticlockwise.

1.4 | Torque

Torques (T; N m) are moments that do not arise from the application of a single force but rather as a twisting effect. An example is the effect of a screwdriver which is rotated not by a single force at a point but rather by a distributed turning motion. Torques are measured in (N m) and can be combined with other moments.

EXAMPLE 1.5

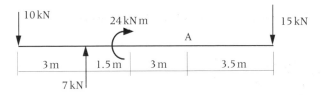

Solution

The clockwise torque, 24 kN m, should be added directly into the moment calculation:

$$M_A = -(10 \times 7.5) + (7 \times 4.5) + 24 + (15 \times 3.5)$$
$$= 33 \text{ kN m}$$

The total moment at A is 33 kN m clockwise.

1.5 Differentiation

The derivative of a function $f(x)$ at $x = a$ is the gradient of the curve of $f(x)$ at that particular value of x. The derivative is defined by the following, where 'lim' refers to the limit as δx tends to zero:

$$y' = \frac{dy}{dx} = \lim \frac{\delta y}{\delta x} = \lim \left[\frac{y(a + \delta x) - y(a)}{\delta x} \right]$$

(1.5)

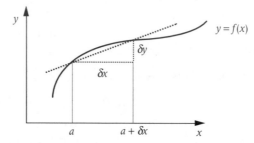

The line through the neighbouring points on $f(x)$ at $x = a$ and $x = a + \delta x$ gets closer to the gradient of the curve at $x = a$, defined by the ratio $\delta y / \delta x$, as δx approaches zero.

Differentiation therefore provides a means of establishing the rate of change of a function. As you will see in the following chapters, rate of change arises in many applications in mechanics. Differentiation can be carried out from first principles as shown in Example 1.6, however this rather cumbersome process is unnecessary if we apply the standard rules which are illustrated by Examples 1.7 to 1.19.

EXAMPLE 1.6

Differentiate the following functions from first principles:

(a) $y = x^2$

$$\frac{\delta y}{\delta x} = \left[\frac{y(a + \delta x) - y(a)}{\delta x} \right]$$

$$= \left[\frac{(x + \delta x)^2 - x^2}{\delta x} \right]$$

$$= \left[\frac{\left(x^2 + 2x\delta x + (\delta x)^2 \right) - x^2}{\delta x} \right]$$

$$= [2x + \delta x]$$

$$\frac{dy}{dx} = \lim \frac{\delta y}{\delta x} = 2x$$

(b) $\quad y = 4x$

$$\frac{\delta y}{\delta x} = \left[\frac{y(a + \delta x) - y(a)}{\delta x} \right]$$

$$= \left[\frac{4(x + \delta x) - 4x}{\delta x} \right]$$

$$= \left[\frac{(4x + 4\delta x) - 4x}{\delta x} \right]$$

$$= 4$$

$$\frac{dy}{dx} = \lim \frac{\delta y}{\delta x} = 4$$

1.5.1   Differentiation rules for algebraic functions

In the tables of rules that follow, c is a constant and u and v are differentiable functions of x:

$$\frac{d}{dx}(x^m) = mx^{m-1} \qquad \text{(a1)}$$

$$\frac{d}{dx}(u^m) = mu^{m-1}\frac{d}{dx}(u) \qquad \text{(a2)}$$

$$\frac{d}{dx}(cu) = c\frac{d}{dx}(u) \qquad \text{(a3)}$$

$$\frac{d}{dx}(uv) = u\frac{d}{dx}(v) + v\frac{d}{dx}(u) \qquad \text{(a4)}$$

$$\frac{d}{dx}\left(\frac{u}{v}\right) = \frac{v\dfrac{d}{dx}(u) - u\dfrac{d}{dx}(v)}{v^2} \qquad \text{(a5)}$$

$$(1.6)$$

Rule (a4) is referred to as the product rule and (a5) is the quotient rule.

In each of the following examples differentiate the given function with respect to x:

EXAMPLE 1.7

$$y = 4 + 2x - 3x^2 - 5x^3 - 8x^4 + 9x^5$$

Solution

This example is straightforward and illustrates the use of the basic rules (a1) and (a3):

$$\frac{dy}{dx} = 0 + 2(1) - 3(2x) - 5(3x^2) - 8(4x^3) + 9(5x^4)$$

$$= 2 - 6x - 15x^2 - 32x^3 + 45x^4$$

EXAMPLE 1.8

$$y = \left(x^2 - 3\right)^4$$

Solution

On inspection, this function follows the pattern described by rule (a2):

$$\frac{dy}{dx} = 4\left(x^2 - 3\right)^3 \cdot \frac{d}{dx}\left(x^2 - 3\right)$$

$$= 4\left(x^2 - 3\right)^3 \cdot (2x)$$

$$= 8x\left(x^2 - 3\right)^3$$

EXAMPLE 1.9

$$y = \left(x^2 + 4\right)^2 \left(2x^3 - 1\right)^3$$

Solution

This function is obviously more complex. It consists of a product, while the individual elements themselves require a bit of care. Take these more complex functions step by step. First use the product rule (a4):

$$\frac{dy}{dx} = \left(x^2 + 4\right)^2 \frac{d}{dx}\left(2x^3 - 1\right)^3 + \left(2x^3 - 1\right)^3 \frac{d}{dx}\left(x^2 + 4\right)^2$$

We can now turn our attention to the next step. The pending differentiations can be carried out using rule (a2):

$$\frac{dy}{dx} = \left(x^2 + 4\right)^2 \cdot 3\left(2x^3 - 1\right)^2 \frac{d}{dx}\left(2x^3 - 1\right) + \left(2x^3 - 1\right)^3 \cdot 2\left(x^2 + 4\right)\frac{d}{dx}\left(x^2 + 4\right)$$

Finally complete the remaining differentiations by using rule (a1):

$$\frac{dy}{dx} = \left(x^2 + 4\right)^2 \cdot 3\left(2x^3 - 1\right) \cdot 6x^2 + \left(2x^3 - 1\right)^3 \cdot 2\left(x^2 + 4\right) \cdot 2x$$

$$= 18x^2\left(x^2 + 4\right)^2\left(2x^3 - 1\right) + 4x\left(2x^3 - 1\right)^3\left(x^2 + 4\right)$$

EXAMPLE 1.10

$$y = \frac{3 - 2x}{3 + 2x}$$

Solution

This example requires the use of the quotient rule (a5) with the resulting pending differentiations completed using rule (a1):

$$\frac{dy}{dx} = \frac{(3+2x)\dfrac{d}{dx}(3-2x)-(3-2x)\dfrac{d}{dx}(3+2x)}{(3+2x)^2}$$

$$= \frac{(3+2x)(-2)-(+3-2x)(2)}{(3+2x)^2}$$

$$= \frac{-6-4x-6+4x}{(3+2x)^2}$$

$$= -\frac{12}{(3+2x)^2}$$

1.5.2 Differentiation rules for trigonometric functions

$$\frac{d}{dx}(\sin u) = \cos u \frac{du}{dx} \qquad \text{(b1)}$$

$$\frac{d}{dx}(\cos u) = -\sin u \frac{du}{dx} \qquad \text{(b2)}$$

$$\frac{d}{dx}(\tan u) = \sec^2 u \frac{du}{dx} \qquad \text{(b3)}$$

$$\frac{d}{dx}(\cot u) = -\csc^2 u \frac{du}{dx} \qquad \text{(b4)}$$

$$\frac{d}{dx}(\sec u) = \sec u \tan u \frac{du}{dx} \qquad \text{(b5)}$$

$$\frac{d}{dx}(\csc u) = -\csc u \cot u \frac{du}{dx} \qquad \text{(b6)}$$

$$(1.7)$$

EXAMPLE 1.11

$$y = \sin(3x) + \cos(2x)$$

Solution

The initial trigonometric differentiations are carried out using rules (b1) and (b2) leaving algebraic expressions which can be differentiated by rule (a1):

$$\frac{dy}{dx} = \cos(3x)\frac{d}{dx}(3x) - \sin(2x)\frac{d}{dx}(2x)$$

$$= 3\cos(3x) - 2\sin(2x)$$

EXAMPLE 1.12

$$y = x^2 \sin(4x)$$

Solution

This example is a combination of product rule (a4), which should be applied first, followed by trigonometric rule (b1) and finally algebraic rule (a1):

$$\frac{dy}{dx} = x^2 \frac{d}{dx}(\sin 4x) + \sin(4x)\frac{d}{dx}(x^2)$$

$$= x^2 \cos(4x)\frac{d}{dx}(4x) + \sin(4x) \cdot 2x$$

$$= 4x^2 \cos(4x) + 2x\sin(4x)$$

EXAMPLE 1.13

$$y = \frac{\cos(5x^2)}{x^2}$$

Solution

This time we should recognise that we are dealing with a quotient. Apply rule (a5) first, then trigonometric rule (b2) and complete the pending algebraic differentiations using rule (a1):

$$\frac{dy}{dx} = \frac{x^2 \frac{d}{dx}(\cos(5x^2)) - \cos(5x^2)\frac{d}{dx}(x^2)}{(x^2)^2}$$

$$= \frac{-x^2 \sin(5x^2)\frac{d}{dx}(5x^2) - 2x\cos(5x^2)}{x^4}$$

$$= \frac{-x^2 \cdot 10x\sin(5x^2) - 2x\cos(5x^2)}{x^4}$$

$$= \frac{-10x^3 \sin(5x^2) - 2x\cos(5x^2)}{x^4}$$

EXAMPLE 1.14

$$y = \tan^3(x^5)$$

Solution

Note the interpretation of the form of this equation in the first line of the solution. The function initially requires us to use rule (a2). The remaining trigonometric functions can then be differentiated using rule (b3):

$$y = \tan^3(x^5)$$

$$y = \left(\tan(x^5)\right)^3$$

$$\frac{dy}{dx} = 3\left(\tan(x^5)\right)^2 \frac{d}{dx}\left(\tan(x^5)\right)$$

$$= 3\tan^2(x^5)\sec^2(x^5)\frac{d}{dx}(x^5)$$

$$= 3\tan^2(x^5)\sec^2(x^5) \cdot 5x^4$$

$$= 15x^4 \tan^2(x^5)\sec^2(x^5)$$

1.5.3   Differentiation rules for logarithmic functions

$$\frac{d}{dx}(\log_a u) = \frac{1}{u}\log_a e \frac{du}{dx} \qquad \text{(c1)}$$

$$\frac{d}{dx}(\ln u) = \frac{1}{u}\frac{du}{dx} \qquad \text{(c2)}$$

$$(1.8)$$

EXAMPLE 1.15

$$y = \log_6(3x^2 - 5)$$

Solution

This is straightforward enough. Use rules (c1) and (a1):

$$\frac{dy}{dx} = \left(\frac{1}{3x^2-5}\right)\log_6 e \frac{d}{dx}(3x^2-5)$$

$$= \left(\frac{1}{3x^2-5}\right)\log_6(e) \cdot 6x$$

$$= \left(\frac{6x}{3x^2-5}\right)\log_6(e)$$

EXAMPLE 1.16

$$y = \ln^3(2x + 3)$$

Solution

Again it is best to start off by rewriting the function. It is thus clear that we should first use algebraic rule (a2) and then complete the differentiation by using rule (c2):

$$y = \ln^3(2x+3)$$

$$y = (\ln(2x+3))^3$$

$$\frac{dy}{dx} = 3(\ln(2x+3))^2 \frac{d}{dx}(\ln(2x+3))$$

$$= 3\ln^2(2x+3) \cdot \frac{1}{(2x+3)} \frac{d}{dx}(2x+3)$$

$$= 3\ln^2(2x+3) \cdot \frac{1}{(2x+3)} \cdot 2$$

$$= \frac{6\ln^2(2x+3)}{(2x+3)}$$

EXAMPLE 1.17

$$y = \ln(\sin(3x))$$

Solution

First use rule (c2) to deal with the natural logarithm and then apply trigonometric rule (b1):

$$\frac{dy}{dx} = \frac{1}{\sin(3x)} \frac{d}{dx}(\sin(3x))$$

$$= \frac{\cos(3x)}{\sin(3x)} \frac{d}{dx}(3x)$$

$$= \frac{3\cos(3x)}{\sin(3x)}$$

$$= 3\cot(3x)$$

1.5.4 Differentiation rules for exponential functions

$$\frac{d}{dx}(e^u) = e^u \frac{du}{dx} \qquad (e1)$$

$$\frac{d}{dx}(a^u) = a^u \ln a \frac{du}{dx} \qquad (e2)$$

$$(1.9)$$

EXAMPLE 1.18

$$y = e^{x^2}$$

Solution

First apply exponential rule (e1) and then (a1) to the pending differentiation of the algebraic function:

$$\frac{dy}{dx} = e^{x^2}\frac{d}{dx}(x^2)$$

$$= 2xe^{x^2}$$

EXAMPLE 1.19

$$y = e^{-2x}\sin(3x)$$

Solution

Deal first with the fact that this example is a product by using rule (a4). We are then left with a trigonometric function that can be differentiated by rule (b1), and an exponential function that can be differentiated using rule (e1):

$$\frac{dy}{dx} = e^{-2x}\frac{d}{dx}(\sin(3x)) + \sin(3x)\frac{d}{dx}(e^{-2x})$$

$$= e^{-2x}\cos(3x)\frac{d}{dx}(3x) + \sin(3x)\cdot e^{-2x}\frac{d}{dx}(-2x)$$

$$= 3e^{-2x}\cos(3x) - 2e^{-2x}\sin(3x)$$

$$= e^{-2x}(3\cos(3x) - 2\sin(3x))$$

1.5.5 Numerical differentiation

Not all information is presented in neat formulae. Data are often given in numerical tabular format. In these cases numerical differentiation techniques are used to determine the rate of change and a particularly simple method to use is the central difference formula.

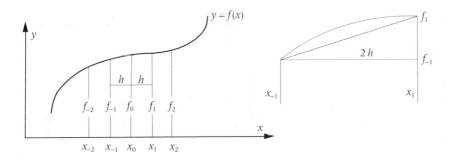

By referring to the diagrams above (the right-hand diagram is an enlargement of the section between x_{-1} and x_1) it can be seen that to determine the rate of change (gradient) at $x = x_0$, we need to divide the difference between the values of the function at x_1 and x_{-1} by twice the interval step, h:

$$f_0' \approx \frac{f_1 - f_{-1}}{2h}$$

(1.10)

EXAMPLE 1.20

The table shows $f(x)$ for $0 \leq x \leq 10$ for a given set of data. Complete the central difference table and hence determine $f'(x)$, $f''(x)$ and $f'''(x)$.

x	$f(x)$
0	6
1	9
2	18
3	39
4	78
5	141
6	234
7	363
8	534
9	753
10	1026

Solution

First complete column 3 of the table starting with $f'(1)$ as shown below, followed by $f'(2)$, and so on. Once column 3 is complete go on to fill in column 4 starting with $f''(2)$. Finally column 5 can be entered starting with $f'''(3)$.

$$f'(1) \approx \frac{18-6}{2(1)} \approx 6 \quad f'(2) \approx \frac{39-9}{2(1)} \approx 15 \quad f''(2) \approx \frac{30-6}{2(1)} \approx 12$$

$$f'''(3) \approx \frac{24-12}{2(1)} \approx 6 \quad \text{etc.}$$

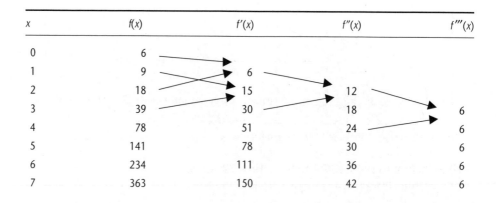

x	$f(x)$	$f'(x)$	$f''(x)$	$f'''(x)$
0	6			
1	9	6		
2	18	15	12	
3	39	30	18	6
4	78	51	24	6
5	141	78	30	6
6	234	111	36	6
7	363	150	42	6

x	$f(x)$	$f'(x)$	$f''(x)$	$f'''(x)$
8	534	195	48	
9	753	246		
10	1026			

The table shows the calculated values of $f'(x)$, $f''(x)$ and $f'''(x)$ at various data points. The pyramidal effect of the process results in the progressive loss of results at the edges of the table.

Note that in fact:

$$f(x) = x^3 + 2x + 6, \quad f'(x) = 3x^2 + 2, \quad f''(x) = 6x \quad \text{and} \quad f'''(x) = 6$$

so

$$f(6) = 6^3 + 2(6) = 234, \quad f'(6) = 3(6)^2 + 2 = 110, \quad f''(6) = 6(6) = 36$$
and $f'''(6) = 6$

The difference table results compare favourably with these.

1.6　Integration

Integration is defined as the reverse process of differentiation. Therefore if $\frac{dy}{dx} = z(x)$ then:

$$y(x) = \int^x z(x)dx \tag{1.11}$$

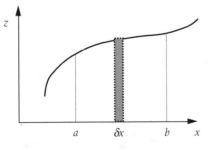

The shaded strip has an approximate area equal to $z(x)\delta x$. The total area between $x = a$ and $x = b$ can be found by adding up all such strips between a and b. As δx for each strip tends to zero this area is given by the integral $\int_a^b z(x)dx$.

Generally, for all but the most straightforward of functions, integration is much trickier than differentiation. However, for our purposes, we need only consider functions which match the types shown below.

1.6.1 Basic rules for integration

$$\int u^m du = \frac{u^{m+1}}{m+1} + C \qquad (f1)$$

$$\int \frac{1}{u} du = \ln|u| + C \qquad (f2)$$

$$\int a^u du = \frac{a^u}{\ln a} + C \qquad (f3)$$

$$\int e^u du = e^u + C \qquad (f4)$$

$$\int \sin(u) \cdot du = -\cos(u) + C \qquad (f5)$$

$$\int \cos(u) \cdot du = \sin(u) + C \qquad (f6)$$

$$\int u \cdot dv = uv - \int v \cdot du \qquad (f7)$$

$$(1.12)$$

Note the constant of integration, C, in each case. Rule (f7) is the 'integration by parts' procedure and is a consequence of the product rule for differentiation (a4):

$$\frac{d}{dx}(uv) = u\frac{d}{dx}(v) + v\frac{d}{dx}(u)$$

$$uv = \int u\frac{d}{dx}(v) \cdot dx + \int v\frac{d}{dx}(u) \cdot dx$$

$$uv = \int u \cdot dv + \int v \cdot du$$

$$\int u \cdot dv = uv - \int v \cdot du$$

EXAMPLE 1.21

$$y = \int (2x^2 - 5x + 3) \cdot dx$$

Solution

This example is typical of the basic algebraic functions that can be integrated directly using rule (f1):

$$y = \int (2x^2 - 5x + 3) \cdot dx$$

$$= \frac{2x^3}{3} - \frac{5x^2}{2} + 3x + C$$

EXAMPLE 1.22

$$y = \int \left(\frac{1}{2x-3}\right) \cdot dx$$

Solution

This function is potentially similar to (f2) but requires some initial modification. Start by investigating u to determine du and then, if possible, amend the initial function to conform to rule (f2):

$$u = 2x - 3$$

$$\frac{du}{dx} = 2$$

$$du = 2 \cdot dx$$

The required amendment is to introduce the factor 2 into the integration. This is permissible provided we cancel out the effect by also dividing, outside the differentiation, by 2:

$$y = \frac{1}{2} \int \left(\frac{2}{2x - 3} \right) dx$$

$$y = \frac{1}{2} \int \frac{1}{u} \cdot du$$

$$y = \frac{1}{2} \ln|2x - 3| + C$$

EXAMPLE 1.23

$$y = \int \cos(3x) \cdot dx$$

Solution

We can use rule (f6) after a slight adjustment to make the function conform to the correct pattern:

$$y = \int \cos(3x) \cdot dx$$

$$= \frac{1}{3} \int \cos(3x) \cdot 3dx$$

$$= \frac{1}{3} \sin(3x) + C$$

EXAMPLE 1.24

$$y = \int e^{4x} \cdot dx$$

Solution

Rule (f4) is applicable after amending the function to conform to the rule:

$$y = \int e^{4x} \cdot dx$$

$$= \frac{1}{4} \int e^{4x} \cdot 4dx$$

$$= \frac{1}{4} e^{4x} + C$$

EXAMPLE 1.25

$$y = \int a^{4x} \cdot dx$$

Solution

Rule (f3) can be applied once the function has been made to conform to the rule:

$$y = \int a^{4x} \cdot dx$$

$$= \frac{1}{4} \int a^{4x} \cdot 4dx$$

$$= \frac{a^{4x}}{4 \ln a}$$

EXAMPLE 1.26

$$y = \int x \sin(x) \cdot dx$$

Solution

This example is tackled using the integration by parts procedure, rule (f7). Begin by assigning u and dv to the original function and hence calculate du and v:

Let $u = x$ and $dv = \sin(x)dx$

$$\frac{du}{dx} = 1 \qquad\qquad v = -\cos(x)$$

$$du = dx$$

Now apply rule (f7). The resulting integration of the trigonometric function is straightforward and uses rule (f6):

$$\int u \cdot dv = uv - \int v \cdot du$$

$$= (x)(-\cos(x)) - \int -\cos(x) \cdot dx$$

$$= -x\cos(x) + \sin(x) + C$$

It should be noted that the original choice for u and dv is critical. In this case the resulting function was easier to integrate than the original. This is not always the case, so don't expect the method to work every time.

EXAMPLE 1.27

$$y = \int xe^x \cdot dx$$

Solution

This is another example of the use of the integration by parts procedure:

Let $u = x$ and $dv = e^x \cdot dx$

$$\frac{du}{dx} = 1 \qquad v = e^x$$

$$\int u \cdot dv = uv - \int v \cdot du$$

$$= xe^x - \int e^x \cdot dx$$

$$= xe^x - e^x + C$$

$$= e^x(x-1) + C$$

1.6.2 Definite integrals

When the upper and lower limits, a and b, of an integration are specified, the value associated with the integration can be calculated by substituting a and b into the resulting integrated equation and subtracting the lower limit value from that of the upper limit. The constant of integration, C, which appears in both cases, consequently cancels out and can be ignored.

EXAMPLE 1.28

Calculate the result of the following integration:

$$y = \int_2^5 (4x^2 - 3x) \cdot dx$$

Solution

Carry out the integration in the normal way and then calculate the result by substituting the upper and lower limits into the resulting equation:

$$y = \int_2^5 (4x^2 - 3x) \cdot dx$$

$$= \left[\left(\frac{4x^3}{3} - \frac{3x^2}{2} \right) \right]_2^5$$

$$= \left[\left(\frac{4(5)^3}{3} - \frac{3(5)^2}{2} \right) - \left(\frac{4(2)^3}{3} - \frac{3(2)^2}{2} \right) \right]$$

$$= (129.167) - (4.667)$$

$$= 124.5$$

1.6.3 Numerical integration

Frequently we wish to integrate a function which is only defined as a tabulation of data. To do this, as was the case with differentiation, we require a numerical process. Two such procedures are the Trapezoidal rule and Simpson's rule.

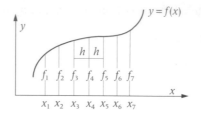

The Trapezoidal rule assumes that the area under the curve $y = f(x)$ can be represented by a series of trapeziums which have a width equal to the step size, h. This introduces an error due to the small areas between the curve and the top edge of the trapeziums, of course.

$$\int_{x_1}^{x_2} f(x) \cdot dx \approx \frac{h(f_1 + f_2)}{2}$$

$$\int_{x_2}^{x_3} f(x) \cdot dx \approx \frac{h(f_2 + f_3)}{2}$$

etc.

Therefore generally for n sub-intervals, according to the Trapezoidal rule, we have:

$$\int_{x_1}^{x_{n+1}} f(x) \cdot dx \approx \frac{h}{2}[f_1 + 2f_2 + 2f_3 + 2f_4 + \ldots + f_{n+1}] \qquad (1.13)$$

Simpson's rule gives a better approximation and this is achieved by fitting the arc of a parabola to the curve at three adjacent tabular points. The rule requires there to be an even number of sub-intervals:

$$\int_{x_1}^{x_{n+1}} f(x) \cdot dx \approx \frac{h}{3}[f_1 + f_{n+1} + 4(f_2 + f_4 + f_6 + \ldots) + 2(f_3 + f_5 + f_7 + \ldots)]$$

$$(1.14)$$

EXAMPLE 1.29

Integrate $f(x)$ between $x = 1$ and $x = 7$ from the data table given below:

x	1	1.5	2	2.5	3	3.5	4	4.5	5	5.5	6	6.5	7
$f(x)$	0	3	7	12	18	25	33	42	52	63	75	88	102

Solution

The solution can be obtained using either of the two rules. First we will calculate an answer using the Trapezoidal rule:

x	1	1.5	2	2.5	3	3.5	4	4.5	5	5.5	6	6.5	7
$f(x)$	0	3	7	12	18	25	33	42	52	63	75	88	102
	f_1	f_2	f_3	f_4	f_5	f_6	f_7	f_8	f_9	f_{10}	f_{11}	f_{12}	f_{13}

$$\int_{x_1}^{x_{n+1}} f(x) \cdot dx \approx \frac{h}{2}[f_1 + 2f_2 + 2f_3 + 2f_4 + \ldots f_{n+1}]$$

$$\approx \frac{0.5}{2}[0 + 2(3 + 7 + 12 + 18 + 25 + 33 + 42 + 52 + 63 + 75 + 88) + 102]$$

$$\approx 234.5$$

There are 13 data points giving 12 sub-intervals. Since there is an even number of sub-intervals we can apply Simpson's rule:

$$\int_{x_1}^{x_{n+1}} f(x) \cdot dx \approx \frac{h}{3}[f_1 + f_{n+1} + 4(f_2 + f_4 + f_6 + \ldots) + 2(f_3 + f_5 + f_7 + \ldots)]$$

$$\approx \frac{0.5}{3}[0 + 102 + 4(3 + 12 + 25 + 42 + 63 + 88) + 2(7 + 18 + 33 + 52 + 75)]$$

$$\approx 234$$

It is interesting to note that the data used in the table were generated from the equation $f(x) = 2x^2 + x - 3$, so we can check the actual value of the integral against those obtained by the two numerical processes:

$$f(x) = 2x^2 + x - 3$$

$$\int f(x) \cdot dx = \frac{2x^3}{3} + \frac{x^2}{2} - 3x + C$$

$$\int_1^7 f(x) \cdot dx = \left[\frac{2x^3}{3} + \frac{x^2}{2} - 3x\right]_1^7$$

$$= \left(\frac{2(7)^3}{3} + \frac{(7)^2}{2} - 3(7)\right) - \left(\frac{2(1)^3}{3} + \frac{(1)^2}{2} - 3(1)\right)$$

$$= 232.167 - (-1.833)$$

$$= 234$$

1.7 | Centre of mass

If a set of forces acts on an object, then the motion of the centre of mass, in translation or rotation, is the same as would be expected if the object were represented by a single particle, concentrated at the centre of mass, having the total mass of the object.

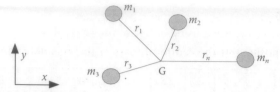

Consider a series of masses, m_1, m_2, m_3, ..., m_n which have position vectors r_1, r_2, r_3, ..., r_n relative to the centre of mass, G, then the position of the centre of mass is established when:

$$\sum_{i=1}^{n} m_i r_i = 0 \qquad (1.15)$$

The symbol $\sum_{i=1}^{n}$ simply means 'add up all the mr^2 terms'. A vector approach to the solution of mechanics problems is not pursued here. It is sufficient to appreciate that the magnitude of the position vector is given by $r_i = \sqrt{x_i^2 + y_i^2}$ where x_i and y_i are the distances measured in the x- and y-directions between mass m_i and the centroid G.

The position of the centre of mass is best established by considering the position of each individual mass relative to two arbitrary x and y reference axes. If these masses, m_1, m_2, m_3 etc. are distances Y_1, Y_2, Y_3 etc. from the reference axis, x, then the position of the centre of mass relative to the reference axis, $\overline{Y}$, can be established as indicated in the formula below.

Similarly, if the masses m_1, m_2, m_3 etc. are distances X_1, X_2, X_3 etc. from the reference axis, y, then the position of the centre of mass relative to the reference axis, $\overline{X}$, can also be established.

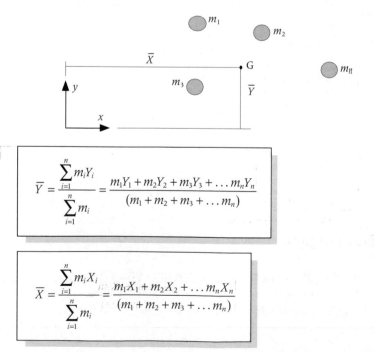

$$\overline{Y} = \frac{\sum_{i=1}^{n} m_i Y_i}{\sum_{i=1}^{n} m_i} = \frac{m_1 Y_1 + m_2 Y_2 + m_3 Y_3 + \ldots m_n Y_n}{(m_1 + m_2 + m_3 + \ldots m_n)}$$

$$\overline{X} = \frac{\sum_{i=1}^{n} m_i X_i}{\sum_{i=1}^{n} m_i} = \frac{m_1 X_1 + m_2 X_2 + \ldots m_n X_n}{(m_1 + m_2 + m_3 + \ldots m_n)} \qquad (1.16)$$

EXAMPLE 1.30

Determine the position of the centre of mass of the assembly shown.

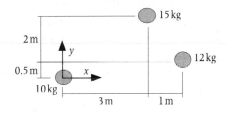

Solution

The most convenient position for the origin of the arbitrary x and y reference axes is at the centre of the 10 kg mass. Use the $\overline{Y}$ and $\overline{X}$ formulae to establish the position of the centre of mass of the assembly relative to the origin of the reference axes.

First determine $\overline{Y}$:

$$\overline{Y} = \frac{\sum\limits_{i=1}^{n} m_i Y_i}{\sum\limits_{i=1}^{n} m_i} = \frac{m_1 Y_1 + m_2 Y_2 + m_3 Y_3 + \ldots m_n Y_n}{(m_1 + m_2 + m_3 + \ldots m_n)}$$

$$\overline{Y} = \frac{(10)(0) + (12)(0.5) + (15)(2.5)}{(10 + 12 + 15)}$$

$$\overline{Y} = 1.176 \text{ m}$$

Next determine $\overline{X}$:

$$\overline{X} = \frac{\sum\limits_{i=1}^{n} m_i X_i}{\sum\limits_{i=1}^{n} m_i} = \frac{m_1 X_1 + m_2 X_2 + \ldots m_n X_n}{(m_1 + m_2 + m_3 + \ldots m_n)}$$

$$\overline{X} = \frac{(10)(0) + (15)(3) + (12)(4)}{(10 + 15 + 12)}$$

$$\overline{X} = 2.514 \text{ m}$$

The position of the centre of mass, G, of the assembly is shown on the diagram overleaf:

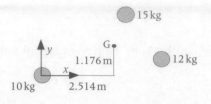

1.8 | Second moment of mass and radius of gyration

The *second moment of mass* (also referred to as the *moment of inertia*), I (kgm^2), is a measure of an object's inertia to angular acceleration; just as mass is a measure of an object's inertia to translational acceleration. The second moment of mass can be calculated from first principles using the following formula but for most practical shapes we can simply use the formulae detailed in Table 1.1. The *radius of gyration*, k (m), is a measure of the compactness of an object and is calculated from the object's mass and second moment of mass.

$$I = \sum_{i=1}^{n} m_i r_i^2 \tag{1.17}$$

$$k = \sqrt{\frac{I}{m}} \tag{1.18}$$

EXAMPLE 1.31

Using Table 1.1 on page 25, determine the second moment of mass and radius of gyration about the *xx*, *yy* and *zz* axes of a solid circular bar of length 0.3 m and diameter 0.1 m. The density of the bar is 1500 kgm^3.

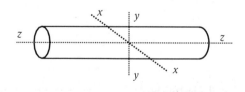

Solution

The second moment of mass of the section, and hence the radius of gyration, about each axis can be established from the formulae in Table 1.1 once the mass of the bar has been calculated:

$$m = \rho V$$
$$= 1500(\pi \times 0.05^2 \times 0.3)$$
$$= 3.534 \text{ kg}$$

$$I_{xx} = I_{yy} = \frac{m(3r^2 + l^2)}{12}$$

$$= \frac{3.534\left(3(0.05)^2 + (0.3)^2\right)}{12}$$

$$= 0.0287 \text{ kg m}^2$$

$$I_{zz} = \frac{mr^2}{2}$$

$$= \frac{(3.534)(0.05)^2}{2}$$

$$= 0.00442 \text{ kg m}^2$$

$$k_{xx} = \sqrt{\frac{I_{xx}}{m}}$$

$$= \sqrt{\frac{0.0287}{3.534}}$$

$$= 0.090 \text{ m}$$

$$k_{yy} = \sqrt{\frac{I_{yy}}{m}}$$

$$= \sqrt{\frac{0.0287}{3.534}}$$

$$= 0.090 \text{ m}$$

$$k_{zz} = \sqrt{\frac{I_{zz}}{m}}$$

$$= \sqrt{\frac{0.00442}{3.534}}$$

$$= 0.035 \text{ m}$$

Table 1.1 Second moment of mass, I, for some simple shapes

(a) *Solid cylinder*

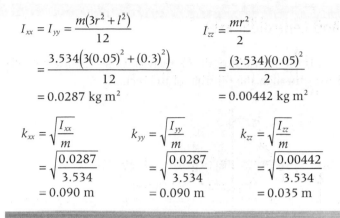

$$I_{xx} = I_{yy} = \frac{m(3r^2 + l^2)}{12}$$

$$I_{zz} = \frac{mr^2}{2}$$

(b) *Thin wheel*

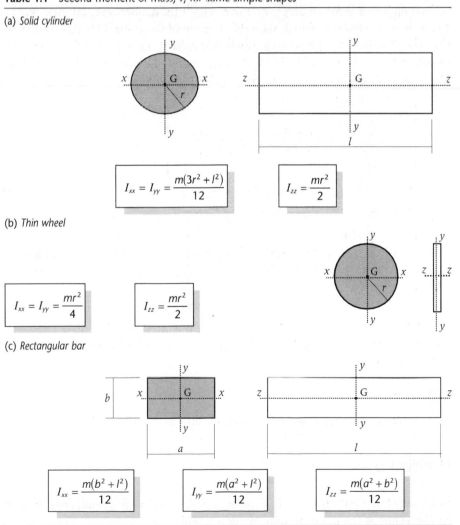

$$I_{xx} = I_{yy} = \frac{mr^2}{4}$$

$$I_{zz} = \frac{mr^2}{2}$$

(c) *Rectangular bar*

$$I_{xx} = \frac{m(b^2 + l^2)}{12}$$

$$I_{yy} = \frac{m(a^2 + l^2)}{12}$$

$$I_{zz} = \frac{m(a^2 + b^2)}{12}$$

1.9 | Centroid and centroidal axes

The *centroid* of a section is the area centre of the section. *Centroidal axes* are any pair of perpendicular axes through the centroid of the section:

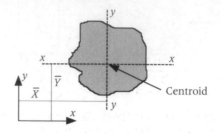

The position of a centroidal axis, such as *xx*, can be determined by dividing the section up into a series of areas, A_1, A_2, A_3 etc. If these areas are distances Y_1, Y_2, Y_3 etc. from a reference axis, *x*, then the position of the centroidal axis relative to the reference axis, $\overline{Y}$, can be established. Similarly if the areas A_1, A_2, A_3 etc. are distances X_1, X_2, X_3 etc. from a reference axis, *y*, then the position of the centroidal axis relative to the reference axis, $\overline{X}$, can also be established.

$$\overline{Y} = \frac{\sum_{i=1}^{n} A_i Y_i}{\sum_{i=1}^{n} A_i} = \frac{A_1 Y_1 + A_2 Y_2 + A_3 Y_3 + \ldots A_n Y_n}{(A_1 + A_2 + A_3 + \ldots A_n)}$$

(1.19)

$$\overline{X} = \frac{\sum_{i=1}^{n} A_i X_i}{\sum_{i=1}^{n} A_i} = \frac{A_1 X_1 + A_2 X_2 + \ldots A_n X_n}{(A_1 + A_2 + A_3 + \ldots A_n)}$$

The position of the centroid is established at the intersection of these two axes. This is very similar to how we established the centre of mass but note that the centroid is established when:

$$\sum_{i=1}^{n} A_i r_i = 0$$

EXAMPLE 1.32

Determine the positions of the centroidal *xx* and *yy* axes of the beam cross-section shown below. The dimensions are shown in mm.

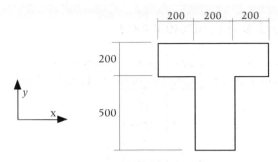

Solution

The section is symmetrical about y so the position of the centroidal yy axis can immediately be determined by inspection. For the centroidal xx axis, a convenient reference line to make measurements from is the bottom of the cross-section. Consider the cross-section as being composed of two rectangular areas, A_1 and A_2, and determine the distance from the horizontal centroidal axis of each to the reference line.

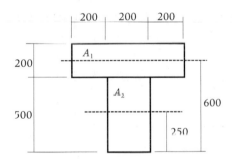

$$A_1 = (200)(600) = 120000 \text{ mm}^2$$
$$A_2 = (200)(500) = 100000 \text{ mm}^2$$
$$A_{total} = 220000 \text{ mm}^2$$

Finally use the $\overline{Y}$ formula which will yield the position of the centroidal axis of the cross-section relative to the reference line:

$$\overline{Y} = \frac{(120000 \times 600) + (100000 \times 250)}{220000}$$

$$= 441 \text{mm}$$

Once the centroidal axis of the cross-section is established relative to the bottom of the section, its position relative to the top of the cross-section and to the local centroidal axes of areas A_1 and A_2 are of course also now known. The diagram overleaf shows these dimensions.

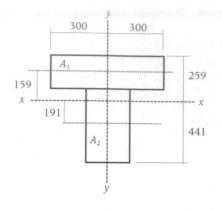

EXAMPLE 1.33

Determine the position of the centroidal *xx* and *yy* axes of the cross-section shown below. The dimensions are shown in mm.

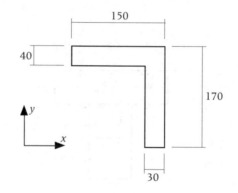

Solution

In this example we will need to calculate the positions for both centroidal axes. First split the section into convenient rectangles. Set the bottom of the section as the reference line. Use the $\bar{Y}$ formula to establish the position of the centroidal *xx* axis relative to this reference line.

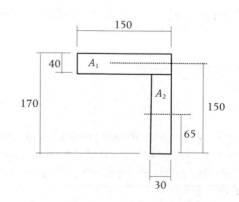

$$A_1 = (150)(40) = 6000\,\text{mm}^2$$
$$A_2 = (130)(30) = 3900\,\text{mm}^2$$
$$A_{\text{total}} = 9900\,\text{mm}^2$$

$$\overline{Y} = \frac{(6000 \times 150) + (3900 \times 65)}{9900}$$
$$= 117\,\text{mm}$$

To determine the position of the centroidal yy axis, set the left edge of the section as the reference line and use the $\overline{X}$ formula:

$$\overline{X} = \frac{(6000 \times 75) + (3900 \times 135)}{9900}$$
$$= 99\,\text{mm}$$

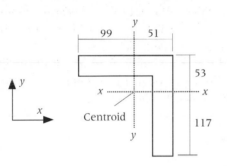

The final diagram shows the positions, rounded to the nearest millimetre, of both axes:

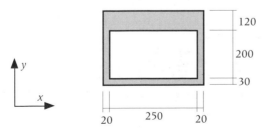

EXAMPLE 1.34

Establish the position of the centroidal xx axis, to the nearest millimetre, of the hollow box section shown below. The dimensions are shown in millimetres.

Solution

The most efficient way to tackle this type of example is to proceed as if the complete section, A_1, was solid, and to consider the hole, A_2, as a negative area. Establish $\overline{Y}$ relative to the bottom of the section.

$$A_1 = (290)(350) = 101\,500\,\text{mm}^2$$
$$A_2 = -(200)(250) = -50\,000\,\text{mm}^2$$
$$A_{\text{total}} = 51\,500\,\text{mm}^2$$

$$\overline{Y} = \frac{(101\,500 \times 175) + (-50000 \times 130)}{51\,500}$$

$$= 219\,\text{mm}$$

The centroidal xx axis is 219 mm above the bottom of the section.

1.10 | Second moment of area about a centroidal axis

The *second moment of area* of an element of a section about an axis in its plane (I_A; m^4) is defined as the product of the section's area and the square of its distance from the axis. Second moments of area have a bearing on the stiffness of a section as will be seen in Chapter 7. Take care not to confuse second moment of area with second moment of mass described in Section 1.8.

A section's second moment of area about a centroidal axis can be determined from first principles using the following formula, but for practical shapes we can simply use the formulae detailed in Table 1.2.

$$I_A = \sum A_i r_i^2 \tag{1.20}$$

Table 1.2 Second moment of area, about a centroidal axis, I_A, of simple sections

(a) *Circle*

$$I_{AXX} = I_{AYY} = \frac{\pi d^4}{64}$$

(b) *Rectangle*

$$I_{AXX} = \frac{ab^3}{12} \qquad I_{AYY} = \frac{ba^3}{12}$$

EXAMPLE 1.35

Determine, from first principles, the second moment of area of a rectangular section about its centroidal xx axis.

Solution

Split the area into thin strips of area dA and determine the second moment of area about the centroidal xx axis by summing the effects of each.

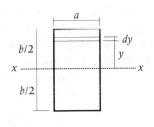

$$I_A = \sum A_i r_i^2$$

Hence the second moment of area about the centroidal xx axis is given by:

$$I_{AXX} = \int y^2 \cdot dA$$

$$= \int_{-b/2}^{b/2} y^2 \cdot a \, dy$$

$$= \int_{-b/2}^{b/2} a y^2 \cdot dy$$

$$= \left[\frac{a y^3}{3} \right]_{-b/2}^{b/2}$$

$$= \left(\frac{a(b/2)^3}{3} \right) - \left(\frac{a(-b/2)^3}{3} \right)$$

$$= \frac{a b^3}{24} - \left(\frac{-a b^3}{24} \right)$$

$$= \frac{a b^3}{12}$$

EXAMPLE 1.36

Determine the second moments of area about the centroidal xx and yy axes of the following sections. Section (b) is a hollow box with a wall thickness of 10 mm. The dimensions shown are in millimetres.

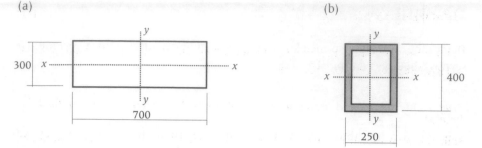

(a)   (b)

Solution

The sections are symmetrical so the positions of the centroidal *xx* and *yy* axes are evident and have been shown on the sections. The second moments of area are calculated using the formulae in Table 1.2. In (b) the section is calculated as if it is solid and then the effect of the hole is deducted.

(a)

$$I_{AXX} = \frac{ab^3}{12}$$

$$= \frac{(700)(300)^3}{12}$$

$$= 1.575 \times 10^9 \, \text{mm}^4$$

$$I_{AYY} = \frac{ba^3}{12}$$

$$= \frac{(300)(700)^3}{12}$$

$$= 8.575 \times 10^9 \, \text{mm}^4$$

(b)

$$I_{AXX} = \frac{(250)(400)^3}{12} - \frac{(230)(380)^3}{12}$$

$$= 1.333 \times 10^9 - 1.052 \times 10^9 \, \text{mm}^4$$

$$= 0.282 \times 10^9 \, \text{mm}^4$$

$$I_{AYY} = \frac{(400)(250)^3}{12} - \frac{(380)(230)^3}{12}$$

$$= 0.521 \times 10^9 - 0.385 \times 10^9 \, \text{mm}^4$$

$$= 0.136 \times 10^9 \, \text{mm}^4$$

1.11 | Parallel axis theorem

We may wish to calculate second moments of area about an axis other than the centroidal axis. This situation occurs when dealing with more complex section shapes. To calculate the second moment of area of a section about a KK axis which is parallel to the section's centroidal *xx* axis and at a distance, *h*, from it, use the parallel axis theorem.

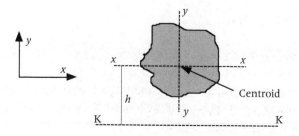

$$I_{AKK} = I_{AXX} + Ah^2 \qquad\qquad (1.21)$$

EXAMPLE 1.37

Determine the second moment of area of the section detailed in Example 1.32.

Solution

The section is redrawn below showing the positions of the centroidal xx and yy axes as calculated previously. The section consists of two rectangles which can be treated separately. The positions of the local centroidal axes of these rectangles relative to the established section centroidal axes are also shown. The contributions of each rectangle to the second moment of area of the section about its centroidal axes are determined using the parallel axis theorem and then summed.

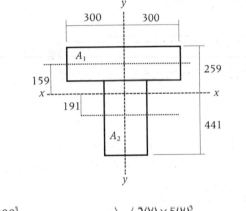

$$I_{AXX} = \left(\frac{600 \times 200^3}{12} + (120000)(159)^2\right) + \left(\frac{200 \times 500^3}{12} + (100000)(-191)^2\right)$$

$$= 9.165 \times 10^9 \, mm^4$$

Note that, in this example, when calculating I_{AYY}, the local centroidal axes of the rectangles coincide with the section centroidal yy axis and therefore $h = 0$.

$$I_{AYY} = \frac{200 \times 600^3}{12} + \frac{500 \times 200^3}{12} = 3.933 \times 10^9 \, mm^4$$

1.12   Direct stress

When a force F acts normal to an element with a cross-sectional area A, then the force is not resisted by just one point on the cross-section. We assume that F is distributed uniformly over the complete face so that there is in fact an intensity of force, referred to as stress. This *direct stress* is given the symbol σ, and is measured in Nm^{-2}.

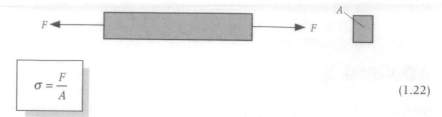

$$\sigma = \frac{F}{A}$$

(1.22)

When the force acts as shown in the diagram it induces tension in the member and is therefore a tensile stress. If instead F were to be pushing onto the member, it would cause a compressive stress.

EXAMPLE 1.38

A hollow circular section tube has an outside diameter of 0.35 m and a wall thickness of 20 mm. If the tube is subjected to a direct compressive force of 150 N, determine the direct stress on the tube.

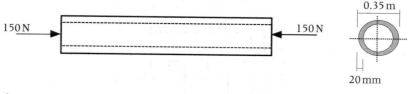

Solution

Assume the force to be evenly distributed over the cross-sectional area. Subtract the area of the hole from the area based on the external diameter:

$$A = \pi \left(\frac{0.35}{2}\right)^2 - \pi \left(\frac{0.35 - 2 \times 0.02}{2}\right)^2$$

$$= 0.0962 - 0.0755$$

$$= 0.0207 \text{ m}^2$$

$$\sigma = \frac{F}{A}$$

$$= \frac{150}{0.0207}$$

$$= 7246 \text{ N m}^{-2}$$

Under the applied load, the tube experiences a direct compressive stress of 7246 N m^{-2}.

1.13 | Direct strain

The application of the tension force F will also cause the member to increase in length, the amount being dependent upon the properties of the material. The application of a compression force would produce a shortening. The *direct strain* on the member, ε, is defined as the change in length per unit length of the member and since both L and δL are measured in m, ε is dimensionless.

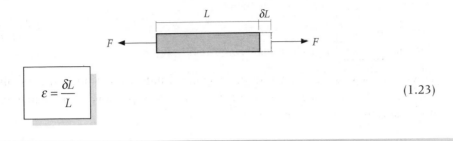

$$\varepsilon = \frac{\delta L}{L}$$ (1.23)

EXAMPLE 1.39

The unloaded length of the tube described in Example 1.38 is 1.3 m. When the 150 N load is applied the length is found to be 1.298 m. Determine the strain experienced by the tube.

Solution

Calculate the change in length of the tube and hence determine the strain.

$$\delta L = 1.3 - 1.298$$
$$= 0.002 \text{ m}$$
$$\varepsilon = \frac{\delta L}{L}$$
$$= \frac{0.002}{1.3}$$
$$= 0.001538$$

Under the applied load, the tube experiences a direct strain of 0.001538.

1.14 | Young's modulus

If we were to plot the applied stress against the resulting strain on a member then the actual shape of the resulting relationship would be dependent upon the material. However, the stress–strain curves for most materials have a number of characteristics in common.

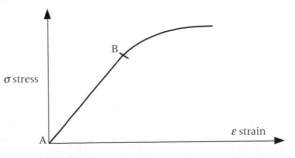

Most importantly, in the initial region, AB, stress is directly proportional to strain and the material behaves elastically. Once the stress is removed the material returns to its original dimensions. This linear relationship is known as *Hooke's law* and can be expressed mathematically by:

$$\sigma = E\varepsilon$$

E is the constant of proportionality and is referred to as *Young's modulus* or the elastic modulus of the material. The units for E are $N\,m^{-2}$. The equation is normally written in the following form:

$$E = \frac{\sigma}{\varepsilon} \tag{1.24}$$

The value of E for a material may or may not be the same in tension and in compression.

EXAMPLE 1.40

A 20 mm diameter wire with a modulus of elasticity of $210\,GN\,m^{-2}$ supports a mass of 5000 kg. The wire has a length of 6 m before the load is attached. Determine the extension of the wire.

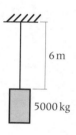

6 m

5000 kg

Solution

First calculate the stress in the wire due to the load. The load should be in N.

$$\sigma = \frac{F}{A}$$

$$\sigma = \frac{(5000)(9.81)}{\pi(0.01)^2}$$

$$\sigma = 156.131 \times 10^6 \ N\,m^{-2}$$

Since the modulus of elasticity of the wire is known, we can determine the corresponding strain. Knowing the strain on the wire and its original length, we can then determine the extension experienced.

$$E = \frac{\sigma}{\varepsilon}$$

$$\varepsilon = \frac{\sigma}{E} \qquad\qquad \varepsilon = \frac{\delta L}{L}$$

$$\varepsilon = \frac{156.131 \times 10^6}{210 \times 10^9} \qquad \delta L = \varepsilon L$$

$$\varepsilon = 0.0007435 \qquad\qquad \delta L = 0.0007435 \times 6$$

$$\delta L = 0.00446 \ m$$

The wire extends by 4.5 mm under the load.

1.15   Further discussion on examples

The examples in this chapter highlight the basic concepts which form the foundation to the topics presented in the chapters which follow.

A sound grasp of the basic rules of calculus is necessary for a proper appreciation of engineering mechanics. For this introductory text you could get by simply with a knowledge of differentiation and integration applied to algebraic functions. However later in your study of mechanics you will find that some problems require the treatment of more complex functions. It is therefore sensible to confront all the rules that you are likely to need now.

It is extremely important to appreciate the difference between second moment of area and second moment of mass. You will find the same symbol, I, used for both in some texts but this can lead to confusion. Here we have used I and I_A. Second moment of mass appears in Chapter 4 when we consider rotational motion. Second moment of area arises in connection with beam stiffness calculations in Chapter 7.

Radius of gyration was defined in Section 1.8 in connection with second moment of mass. However radius of gyration could equally well be defined in terms of second moment of area:

$$k = \sqrt{\frac{I}{m}} \quad \text{where} \quad I = \sum m_i r_i^2$$

$$k = \sqrt{\frac{I_A}{A}} \quad \text{where} \quad I_A = \sum A_i r_i^2$$

The two are equivalent, giving k in metres in each case.

Discussion on stress, strain and elasticity has been kept to the minimum necessary at this stage. Later in your studies you will encounter shear and torsion but these are not a part of this introductory text.

1.16   Problems

1.1 Determine the mass and weight of a circular section concrete column with a diameter of $0.5\,m$ and a height of $7\,m$. The density of the concrete is $2400\,kg\,m^{-3}$.

1.2 In each of the following, determine the total moment at the point A due to the forces shown:

(a)

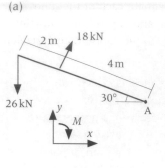

(b)

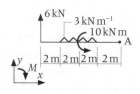

(c)

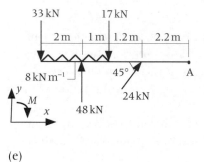

(d)

(e)

(f)

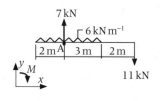

1.3 Differentiate each of the following equations with respect to x:

(a) $y = 18x^2 - 4x + 9$

(b) $y = 12x^3 + \dfrac{12}{x^5}$

(c) $y = (3x + 1)^5$

(d) $y = (2x^2 + 5)^3$

(e) $y = (2x + 5)^{-5}$

(f) $y = \sqrt{3x^2 + x - 1}$

(g) $y = (1 + 3x^2)^{0.3}$

(h) $y = (x^2 - 2x^3)^5$

1.4 Using the product rule, differentiate each of the following equations:

(a) $y = (x^2 - 1)(3x + 2)$

(b) $y = (2 + x)(3 - x)$

(c) $y = (x - 1)(x^2 + x + 1)$

1.5 Using the quotient rule, differentiate each of the following equations with respect to x:

(a) $y = \dfrac{(x + 1)}{(x - 1)}$

(b) $y = \dfrac{(x^3 + 6x)}{(x + 4)}$

(c) $y = \dfrac{4x^3 - 9}{(x - 1)^3}$

(d) $y = \dfrac{x(x + 1)}{3x(x^2 - 5)}$

1.6 Differentiate each of the following trigonometric functions with respect to the appropriate variable:

(a) $y = 4\sin(x)$

(b) $y = \sin(4x)$

(c) $y = \theta - 2\cos(\theta)$

(d) $y = \sin(3x + 2)$

(e) $y = 3\sin(2x) - 4\cos(2 - x)$

(f) $y = 2\sin\left(3x - \dfrac{\pi}{2}\right)$

(g) $y = \cos(1 - 2\pi t)$

(h) $y = 3\sin^3(x^2 - 6x)$

1.7 Differentiate each of the following functions:

(a) $y = 3\theta\sin\theta$

(b) $y = z^2\cos(z)$

(c) $y = (t^2 + 1)\sin(2t)$

(d) $y = x^3\cos^2 x$

(e) $y = \dfrac{3\sin(t)}{t^4}$

(f) $y = \dfrac{\sin(\theta)}{(1 + \theta)}$

(g) $y = \dfrac{\sin(x)}{(1 + \cos(x))}$

(h) $y = \dfrac{\sin(3\theta)}{\cos(2\theta)}$

1.8 Differentiate each of the following exponential and logarithmic functions with respect to the appropriate variable:

(a) $y = e^{3x}$

(b) $y = e^{2x+1}$

(c) $y = e^{(3x^2+4x)}$

(d) $y = xe^x$

(e) $y = \dfrac{e^x}{x+1}$

(f) $y = e^{2\theta}\cos(\theta)$

(g) $y = e^{2x}\sin(3x)$

(h) $y = \dfrac{\cos(x)}{e^x}$

(i) $y = e^{\sin(x)\cos(x)}$

(j) $y = \ln(3x)$

(k) $y = \log_e(1 + x)$

(l) $y = 2\ln(2 + 3x)$

(m) $y = \ln(x^2)$

(n) $y = \ln(\sin(\theta))$

(o) $y = \ln(\sin^2(\theta))$

(p) $y = \ln\left(\dfrac{x}{x+1}\right)$

(q) $y = 2\ln^3 t$

(r) $y = \ln\left(\dfrac{1 + \sin(x)}{1 - \sin(x)}\right)$

1.9 Complete each of the following central difference tables:

(a)

x	$f(x)$	$f'(x)$	$f''(x)$	$f'''(x)$
3	14			
3.1	17.252			
3.2	20.816			
3.3	24.704			
3.4	28.928			
3.5	33.5			
3.6	38.432			
3.7	43.736			
3.8	49.424			
3.9	55.508			
4	62			

(b)

x	$f(x)$	$f'(x)$	$f''(x)$	$f'''(x)$
3	6.3504			
3.1	1.998			
3.2	−2.9888			
3.3	−8.5893			
3.4	−14.7703			
3.5	−21.4855			
3.6	−28.6753			
3.7	−36.2673			
3.8	−44.1761			
3.9	−52.3046			
4	−60.5442			

(c)

x	$f(x)$	$f'(x)$	$f''(x)$	$f'''(x)$
3	20.0855			
3.1	22.1979			
3.2	24.5325			
3.3	27.1126			
3.4	29.964			
3.5	33.1154			
3.6	36.5982			
3.7	40.4472			
3.8	44.7011			
3.9	49.4023			
4	54.598			

1.10 Integrate the following functions:

(a) $y = \int (4x^3 - 3x^2 + 6) \cdot dx$

(b) $y = \int (3x^2 + 2x^2) \cdot dx$

(c) $y = \int \left(\frac{4x^3}{3} \right) \cdot dx$

(d) $y = \int \left(-\frac{3x^2}{5} \right) \cdot dx$

(e) $y = \int \left(2x - \frac{5}{x^3} + \frac{4}{x^2} + \frac{2}{x} \right) \cdot dx$

1.11 Adopt appropriate substitutions and carry out the following integrations:

(a) $y = \int \frac{(x+3)}{(x^2+6x)} \cdot dx$

(b) $y = \int 5e^{3x} \cdot dx$

(c) $y = \int 4\sin(3x) \cdot dx$

1.12 Carry out the following integration using the integration by parts procedure:

$$y = \int x \cos(x) \cdot dx$$

1.13 Calculate the result of the following definite integrations:

(a) $y = \int_4^6 (3x^2 + 10) \cdot dx$

(b) $y = \int_{-3}^3 (6x^3 - 4x) \cdot dx$

(c) $y = \int_2^7 (4x^2 - 9x) \cdot dx$

1.14 The three tables below detail the functions in **1.13** numerically. Integrate the functions between the limits given in **1.13** using a numerical integration procedure with the step size as given in the tables, and compare the results obtained with those in **1.13**, explaining any differences that occur.

(a)

x	4	4.2	4.4	4.6	4.8	5	5.2	5.4	5.6	5.8	6
$f(x)$	58	62.92	68.08	73.48	79.12	85	91.12	97.48	104.08	110.92	118

(b)

x	−3	−2.5	−2	−1.5	−1	−0.5	0	0.5	1	1.5	2	2.5	3
$f(x)$	−150	−83.75	−40	−14.25	−2	1.25	0	−1.25	2	14.25	40	83.75	150

(c)

x	2	2.5	3	3.5	4	4.5	5	5.5	6	6.5	7
$f(x)$	−2	2.5	9	17.5	28	40.5	55	71.5	90	110.5	133

1.15 Determine the position of the centre of mass of the assembly shown relative to the 3 kg mass:

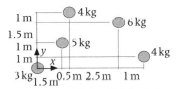

1.16 Using Table 1.1, determine the second moment of mass and radius of gyration about the xx, yy and zz axes of the solid rectangular bar shown. The bar has a length of 0.45 m and a density of 1850 kg m³.

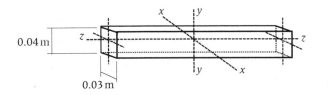

1.17
(a) Determine the position of the centroidal xx and yy axes of the cross-section shown overleaf.
(b) Calculate the second moment of area of the cross-section about the xx axis.
(c) Calculate the second moment of area of the cross-section about the yy axis.

The section is symmetrical about yy and all dimensions are in mm.

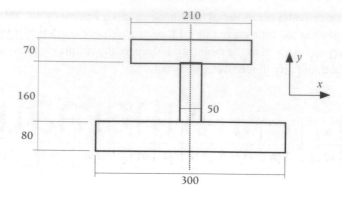

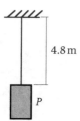

4.8 m

P

1.18 A 4.8 m long, 16 mm diameter wire with a modulus of elasticity of 200 GN m^{-2} extends 6 mm when a mass of P kg is hung from one end. Determine the magnitude of the mass P.

CHAPTER 2

Kinematics

Kinematics is the study and analysis of the motion of bodies without consideration of the forces, moments etc. that produced these motions. Thus kinematics is concerned purely with the relationships between position, velocity and acceleration. Strictly position, velocity and acceleration are vectors and therefore have magnitude and direction. Occasionally the term 'speed' is used in place of velocity. Speed is a scalar and has magnitude only.

In this chapter we will consider basic linear and rotational motion. In particular we will develop and use the standard constant acceleration models.

The following concepts are introduced in this chapter:

- Position
- Velocity
- Acceleration

- Angular displacement
- Angular velocity
- Angular acceleration

2.1 | The position/velocity/acceleration relationship

Velocity is defined as the rate of change of position and *acceleration* as the rate of change of velocity. The following basic equations therefore link the three concepts, where x (m), v (ms^{-1}) and a (ms^{-2}) define position, velocity and acceleration respectively.

$$x$$
$$\frac{dx}{dt} = v$$
$$\frac{dv}{dt} = a$$

(2.1)

$$a$$
$$\int a\,dt = v$$
$$\int v\,dt = x$$

(2.2)

EXAMPLE 2.1

The burning of a rocket flare gives a vertical velocity which can be expressed by:

$$v = 27t^2 - 6t^3$$

where v is given in ms^{-1} and time t in seconds.

The rocket is fired from ground level so that $x = 0$ at $t = 0$.

(a) Determine expressions for the acceleration and height of the rocket.

(b) Sketch the motion of the rocket.

(c) Determine the maximum velocity of the rocket.

(d) Calculate the maximum height reached by the rocket.

Solution

(a) The equation describing the acceleration of the rocket as a function of time t can be obtained by differentiating the given expression for v:

$$a = \frac{dv}{dt}$$
$$a = (54t - 18t^2)\,\text{ms}^{-2}$$

To obtain the required equation for position (in this case the height reached above ground level) we need to integrate the given expression for v. This introduces a constant of integration, C.

$$x = \int v\,dt$$
$$x = 9t^3 - 1.5t^4 + C$$

To eliminate the constant of integration we substitute a known solution into the equation and solve for C. In this case we know that $x = 0$ at $t = 0$:

$$0 = 9(0)^3 - 1.5(0)^4 + C$$
$$C = 0$$

Hence $x = (9t^3 - 1.5t^4)\,\text{m}$

(b) A table of x, v and a against t can be produced by substituting values for t between 0 and 6 s for, say, 1 s increments. The results obtained were used to produce the graph shown at the end of this example.

t (s)	x (m) = $9t^3 - 1.5t^4$	v (ms^{-1}) = $27t^2 - 6t^3$	a (ms^{-2}) = $54t - 18t^2$
0	0	0	0
1	7.5	21	36
2	48	60	36
3	121.5	81	0
4	192	48	−72
5	187.5	−75	−180
6	0	−324	−324

(c) The time when the velocity of the rocket is at a maximum occurs when $\dfrac{dv}{dt} = 0$, i.e. when the acceleration $a = 0$. Solving the resulting equation for t and substituting this value back into the equation for v yields the maximum velocity, v_{max}.

$$0 = 54t - 18t^2$$
$$t(54 - 18t) = 0$$

Hence $t = 0\,\text{s}$ or $3\,\text{s}$

$$v = 27t^2 - 6t^3$$
$$\therefore v_{max} = 27(3)^2 - 6(3)^3$$
$$v_{max} = 81\,\text{m s}^{-1}$$

The maximum velocity of the rocket is $81\,\text{m s}^{-1}$ and occurs after 3 s.

(d) The time when the rocket reaches its maximum height occurs when $\dfrac{dv}{dt} = 0$, i.e. when the velocity $v = 0$. Solving the resulting equation for t and substituting the value found in this case back into the equation for x yields the maximum height, x_{max}.

$$0 = 27t^2 - 6t^3$$

$$t^2(27 - 6t) = 0$$

Hence $t = 0\,\text{s}$ or $t = 4.5\,\text{s}$

$$x = 9t^3 - 1.5t^4$$

$$\therefore x_{max} = 9(4.5)^3 - 1.5(4.5)^4$$

$$x_{max} = 205.031\,\text{m}$$

The maximum height reached by the rocket is 205.031 m and this occurs after 4.5 s.

Presenting the results obtained for the position, velocity and acceleration of the rocket during the time 0 s to 6 s on a single graph, as shown below, gives a useful picture of the motion.

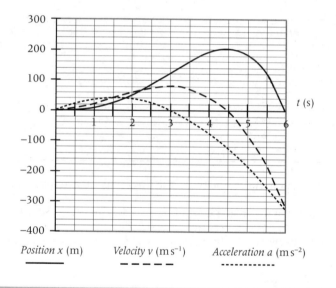

Position x (m)   Velocity v (m s^{-1})   Acceleration a (m s^{-2})

EXAMPLE 2.2

Car A starts a journey at 10.00 am travelling at a constant speed of 45 km h^{-1}. Car B begins the same journey at 10.15 am travelling at a constant speed of 65 km h^{-1}.

(a) At what time does car B overtake car A?

(b) How far have the cars travelled when B overtakes A?

Solution

Create equations to describe the position of each car with respect to time.

(a) Let x_A and x_B represent the positions of cars A and B respectively and let v_A and v_B be their speeds.

$$x_A = \int v_A dt \qquad x_B = \int v_B dt$$
$$= \int \left(\frac{45 \times 10^3}{60^2}\right) dt \qquad = \int \left(\frac{65 \times 10^3}{60^2}\right)$$
$$= 12.5t + C_1 \qquad = 18.0556t + C_2$$

To determine the constants of integration, C_1 and C_2, substitute into the equations the known solutions that $x_A = 0$ at $t = 0$ and $x_B = 0$ at $t = 900$ s:

$$0 = 12.5(0) + C_1 \qquad 0 = 18.0556(900) + C_2$$
$$C_1 = 0 \qquad C_2 = -16250$$

So the equation can be written as:

$$x_A = 12.5t \quad \text{(eqn 1)} \quad \text{and} \quad x_B = 18.0556t - 16250 \quad \text{(eqn 2)}$$

At the point where car B is alongside car A, x_A and x_B must be equal and hence:

$$12.5t = 18.0556t - 16250$$
$$t = 2925 \text{ s}$$

Car B overtakes car A after 2925 s, i.e. at 10.48 (and 45 s) am.

(b) Substituting 2925 s into either eqn (1) or (2) will determine the distance that the cars have travelled when B overtakes A.

$$x_A = 12.5t$$
$$= 12.5(2925)$$
$$= 36562.5 \text{ m}$$
$$x_B = 18.0556t - 16250$$
$$= 18.0556(2925) - 16250$$
$$= 36562.5 \text{ m}$$

The cars have travelled 36562.5 m when B overtakes A.

2.2 | The constant acceleration model

In Section 2.1 the general relationships between position, velocity and acceleration were introduced. In many practical situations the acceleration is constant, and under this special condition, the following simple set of equations can be applied, where

t = time (s)
u = velocity at time $t = 0$
v = velocity (m s^{-1})

x = displacement (m)

a = acceleration ($\mathrm{m\,s^{-2}}$) constant.

1. A relationship linking initial velocity, final velocity, acceleration and time.

$$\frac{\mathrm{d}v}{\mathrm{d}t} = a$$

$$v = \int (a)\mathrm{d}t$$

$$= at + C_1$$

and since $v = u$ at $t = 0$ substitution gives:

$$u = (a)(0) + C_1$$

$$\therefore C_1 = u$$

$$\boxed{\therefore \ v = u + at}$$

(2.3)

2. A relationship linking position, initial velocity, acceleration and time.

$$x = \int (v)\mathrm{d}t$$

$$= \int (u + at)\mathrm{d}t$$

$$= ut + \frac{1}{2}at^2 + C_2$$

but $x = 0$ at $t = 0$ and substitution gives:

$$0 = (u)(0) + \frac{1}{2}(a)(0)^2 + C_2$$

$$\therefore C_2 = 0$$

$$\boxed{x = ut + \frac{1}{2}at^2}$$

(2.4)

3. A relationship linking position, initial velocity, final velocity and time.

$$v = u + at$$

$$at = (v - u)$$

$$x = ut + \frac{1}{2}at^2$$

$$\therefore x = ut + \frac{1}{2}(v - u)t$$

$$\boxed{x = \frac{1}{2}(u + v)t}$$

(2.5)

4. A relationship linking position, initial velocity, final velocity and acceleration.

$$v = u + at$$
$$v^2 = (u + at)^2$$
$$= u^2 + 2uat + a^2t^2$$
$$= u^2 + a(2ut + at^2)$$
$$2ut + at^2 = 2x$$

$$\boxed{v^2 = u^2 + 2ax}$$ (2.6)

EXAMPLE 2.3

A ball is dropped from the top of a building and takes 2.4 s to reach the ground. Determine the height of the building and the velocity of the ball on striking the ground. Neglect the effects of air resistance.

Solution

This is clearly a constant acceleration problem. The ball accelerates to Earth under the effect of gravity. To solve this type of problem, first set down the available data and then choose the appropriate constant acceleration equation which links the known data to the quantity to be determined.

We have: $u = 0$, $t = 2.4$ s, $a = 9.81\,\mathrm{m\,s^{-2}}$ and require x, hence the appropriate equation is:

$$x = ut + \frac{1}{2}at^2$$
$$= (0)(2.4) + \frac{(9.81)(2.4)^2}{2}$$
$$= 28.25\,\mathrm{m}$$

The building has a height of 28.25 m.

To determine the velocity of the ball on striking the ground, again set out the known data. We now have: $u = 0$, $t = 2.4$ s, $a = 9.81\,\mathrm{m\,s^{-2}}$, $x = 28.25$ m and require v. The appropriate equation this time is:

$$v = u + at$$
$$= (0) + (9.81)(2.4)$$
$$= 23.45\,\mathrm{m\,s^{-1}}$$

The ball strikes the ground with a velocity of $23.54\,\mathrm{m\,s^{-1}}$.

EXAMPLE 2.4

A cricket ball is thrown vertically upwards and takes 3.8 s to return to the ground. Neglecting the height of the person throwing the ball, estimate the throwing speed.

Solution

First set out what we know and what we need to determine. Clearly, if we neglect the height of the thrower then the distance travelled upwards equals that travelled downwards, and as a consequence the time travelling upwards and that travelling downwards are also equal. Using the subscripts 1 and 2 for upward and downward motion respectively we have:

$$x_1 = u_1 t_1 + \frac{1}{2} a_1 t_1^2$$

$$x_2 = u_2 t_2 + \frac{1}{2} a_2 t_2^2$$

$u_2 = 0$, $a_1 = -9.81\,\mathrm{m\,s^{-2}}$, $a_2 = 9.81\,\mathrm{m\,s^{-2}}$, $t_1 = t_2 = 1.9\,\mathrm{s}$ and we require u_1.

$$x_1 = x_2$$

$$u_1 t_1 + \frac{1}{2} a_1 t_1^2 = u_2 t_2 + \frac{1}{2} a_2 t_2^2$$

$$u_1(1.9) - \frac{(9.81)(1.9)^2}{2} = (0)(1.9) + \frac{(9.81)(1.9)^2}{2}$$

$$u_1 = \frac{(9.81)(1.9)^2}{1.9}$$

$$= 18.64\,\mathrm{m\,s^{-1}}$$

Hence the throwing velocity must have been $18.64\,\mathrm{m\,s^{-1}}$.

EXAMPLE 2.5

In this example we will consider free motion in two dimensions. The approach is first to resolve the motion into components in the vertical y and horizontal x axes, and then to consider these separately while maintaining compatibility. In the vertical direction the acceleration experienced is the gravity constant g. In the horizontal direction we can assume that the velocity is constant, i.e. that there is no acceleration. This assumption of course ignores the effect of wind resistance.

A diver jumps off the end of a board with an initial velocity as indicated in the sketch. Calculate the distance out from the end of the diving board to the point where the diver entered the water, and the velocity of the diver on entering the water.

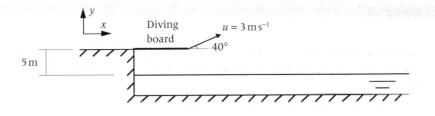

Solution

By considering the vertical motion of the diver, determine the time that the diver is in the air before hitting the water. Apply this time to the horizontal motion to calculate the horizontal distance travelled.

Resolving the diver's velocity into horizontal and vertical components:

$$u_x = 3\cos 40°$$
$$u_y = 3\sin 40°$$

Consider the vertical motion component and determine the time to reach the top of the dive. The acceleration will be due to gravity and will be negative.

$$v_y = u_y + at$$
$$0 = 3\sin 40° - 9.81t$$
$$t = \frac{3\sin 40°}{9.81}$$
$$= 0.20\,\text{s}$$

Determine the vertical distance travelled to the top of the dive:

$$Y = \frac{(u_y + v_y)t}{2}$$
$$= \frac{(3\sin 40° + 0)(0.20)}{2}$$
$$= 0.19\,\text{m}$$

Hence the vertical length of the downwards portion of the dive is 5.19 m. Now calculate the time travelling downwards. This time the acceleration due to gravity is positive.

$$Y = u_y t + \frac{1}{2}at^2$$
$$5.19 = (0)t + \frac{9.81t^2}{2}$$
$$t = \sqrt{\frac{2 \times 5.190}{9.81}}$$
$$= 1.03\,\text{s}$$

Hence the total time in the air before hitting the water = 0.20 + 1.03 = 1.23 s.

Now consider the horizontal motion. The velocity is constant in this direction and the horizontal and vertical motion times must be equal.

$$x = u_x t$$
$$= (3\cos 40°)(1.23)$$
$$= (2.30)(1.23)$$
$$= 2.82\,\text{m}$$

Hence the diver hits the water 2.82 m out from the end of the diving board.

The diver's horizontal velocity is constant at 2.30 m s^{-1} and we can calculate his vertical velocity on hitting the water by considering the vertical component of his downward motion.

$$v_y^2 = u_y^2 + 2ay$$
$$v_y = \sqrt{(0)^2 + (2)(9.81)(1.03)}$$
$$= 10.09\,\text{m s}^{-1}$$

Finally resolve the horizontal and vertical components of velocity to determine the total velocity.

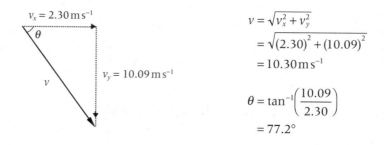

$$v = \sqrt{v_x^2 + v_y^2}$$
$$= \sqrt{(2.30)^2 + (10.09)^2}$$
$$= 10.30\,\text{m s}^{-1}$$

$$\theta = \tan^{-1}\left(\frac{10.09}{2.30}\right)$$
$$= 77.2°$$

Hence the diver enters the water with a velocity of 10.30 m s^{-1} and at an angle of 77.2° to the horizontal.

2.3 | Constant rotational acceleration

The four equations in the previous section are valid for linear motion with constant acceleration. A similar set of equations can be derived for rotational motion. In these equations, linear displacement, velocity and acceleration are replaced by angular displacement, velocity and acceleration, where:

t = time (s)
θ = angular displacement (rad)
ω_1 = initial angular velocity (rad s^{-1})
ω_2 = final angular velocity (rad s^{-1})
α = angular acceleration (rad s^{-2}).

$$\omega_2 = \omega_1 + \alpha t$$

$$\theta = \omega_1 t + \frac{1}{2}\alpha t^2$$

$$\theta = \frac{1}{2}(\omega_1 + \omega_2)t$$

$$\omega_2^2 = \omega_1^2 + 2\alpha\theta \qquad\qquad (2.7)$$

EXAMPLE 2.6

A wheel rotating about its axle accelerates from $10\,\mathrm{rad\,s^{-1}}$ anticlockwise to $15\,\mathrm{rad\,s^{-1}}$ anticlockwise in 3 seconds. Determine the angular acceleration of the wheel and the number of revolutions during the acceleration period.

Solution

Use the rotational form of the kinematic equations. We can determine the angular acceleration from the following equation in which anticlockwise motion has been set as positive:

$$\omega_2 = \omega_1 + \alpha t$$
$$15 = 10 + 3\alpha$$
$$\alpha = 1.667\,\mathrm{rad\,s^{-2}}$$

The positive answer means that the angular acceleration is also anticlockwise. Now we can determine the rotation that takes place, firstly in radians which can then be converted into degrees and finally revolutions.

$$\theta = \omega_1 t + \frac{\alpha t^2}{2}$$

$$= (10)(3) + \frac{(1.667)(3)^2}{2}$$

$$= 37.502\,\mathrm{rad}$$

$$\therefore \theta = 37.502\frac{180°}{\pi}$$

$$= 2148.706°$$

Hence, since there are 360° in one complete revolution of the wheel, the number of revolutions that take place during the acceleration process is given by

$$\frac{2148.706}{360} = 5.969.$$

2.4 | Further discussion on examples

The first two examples in this chapter illustrate the kinematic relationships between position, velocity and acceleration. These are applicable even if the acceleration term is variable. In fact we could extend the sequence position/velocity/acceleration further and determine the rate of change of acceleration, or jerk, J (m s^{-3}):

$$\frac{\mathrm{d}a}{\mathrm{d}t} = J \qquad \int J \cdot \mathrm{d}t = a$$

We could continue further and determine the rate of change of jerk, and so on.

Constant acceleration problems are very common in practice where the motion of a body is affected by gravity. In fact, as you will see in the next chapter, constant acceleration occurs whenever the forces acting on a system are themselves constant. In the gravity examples, the only force acting was the object's weight which is of course constant. Throughout this book the vast majority of situations that we will consider involve forces which do not vary with time and hence we will be dealing with constant acceleration models. Only in the special case of spring forces and strain energy, which are introduced in Chapter 5, where force is proportional to the varying deformation, will we encounter non-constant acceleration.

Examples 2.3 and 2.4, in this chapter, show how the basic equations for constant acceleration can be manipulated to provide the solution to simple one-dimensional problems. Example 2.5 extends the concept into problems defined in two dimensions. In these a common approach is to assume a constant acceleration model vertically and a zero acceleration or constant velocity model horizontally. The horizontal motion ignores the effect of aerodynamic drag. This can in fact be a very important factor which you will no doubt return to later in your studies. Basically, this resistance force is dependent upon the density of the fluid through which the object is travelling, the object's velocity relative to the fluid and a drag coefficient which is established from a dimensionless property of the fluid as it flows, known as the Reynold's Number.

2.5 | Problems

2.1 The figure shows an object moving from left to right along a level track. The position of the object relative to point A, at any time t (s), $0 \leq t \leq 5$ is given by:

$$x = 3t^3 - 12t^2 + 18t$$

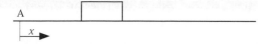

(a) Determine the position of the object relative to A for $t = 0$ to $t = 5$ s in steps of 1 s, and draw a graph of the results.

(b) Calculate the velocity of the object for $t = 0$ to $t = 5$ s in steps of 1 s and superimpose the results on the previous graph.

(c) Calculate the minimum velocity of the object and the time when the minimum velocity occurs.

(d) Determine the acceleration when $t = 5$ s.

2.2 Car A starts a journey 20 minutes before car B. Both cars are travelling at constant speeds which are 60 and 85 km h^{-1} respectively. How long after starting out does car B overtake car A?

2.3 A ball is thrown vertically downwards from a 25 m high building and takes 1.9 s to hit the ground. Determine the initial throwing velocity of the ball and the velocity with which it strikes the ground.

2.4 A balloon is rising with a constant velocity of 2 m s^{-1} when a bag of sand is released. If the bag takes 4.7 s to reach the ground determine:

(a) the height above the ground of the balloon when the bag was released

(b) the height above the ground of the balloon when the bag reaches the ground.

2.5 A train travelling along a track slows down with a constant deceleration from a speed of 60 km h^{-1} to 45 km h^{-1} in 5 s. Determine the deceleration and the distance covered by the train during the deceleration period.

2.6 A train, 45 m long, is standing at rest at a platform when given the all-clear by a signal 140 m ahead of the driver's cabin. The train accelerates uniformly at 0.3 m s^{-2} away from the platform. Determine:

(a) the velocity of the train when the driver's cabin passes the signal

(b) the velocity when the rear of the train passes the signal

(c) the time that passes between the front and rear of the train passing the signal.

2.7 A golfer pitches a ball to a raised green from the point shown on the diagram below. If the initial velocity of the ball is 95 km h^{-1} at angle 60°, how far beyond the hole does the ball pitch on the green? Neglect aerodynamic effects.

2.8 A wheel, rotating about its axle, accelerates at a constant value of $6\,\mathrm{rad\,s^{-2}}$. If the initial angular velocity was $8\,\mathrm{rad\,s^{-1}}$, determine its angular velocity after 4 complete revolutions. Also determine the time that elapsed during these 4 revolutions.

CHAPTER 3

Newton's Laws of Motion: Translational Motion

In this chapter we will consider the solution of problems which can be categorised as purely translational. The elements of the system remain in the same relative position to the mass centre throughout, with no rotation of the system about the mass centre taking place. Chapter 4 is concerned with problems involving fixed-axis rotation and combined rotation and translation.

The chapter first outlines Newton's three laws of motion and then illustrates the application of the second law in problems relating to blocks sliding and accelerating along planes through the application of external forces. Combined systems consisting of two or more connected translating blocks are also introduced. Throughout the chapter the advantages of drawing free-body diagrams are emphasised. The special case of systems in equilibrium is developed, leading into the solution of statics problems.

The following concepts are introduced in this chapter:

- *Newton's laws of motion*
- *Free-body diagrams*
- *Force*
- *Mass*
- *Translational acceleration*
- *Friction*

- *Resolution of forces*
- *Solution of simultaneous equations*
- *Equilibrium*
- *Pin-jointed frames*
- *Struts and ties*

3.1 Newton's laws of motion

The fundamentals of solid mechanics are developed from the initial concepts of force and acceleration as expressed by Newton's three laws of motion. These three laws can be stated in modern terms as:

1st: A system remains at rest, or continues to move in a straight line with constant velocity, if there is no unbalanced force acting on it.

2nd: If there is an unbalanced force acting on a system, the system experiences an acceleration in the direction of the force, proportional to the force and inversely proportional to the mass of the body.

3rd: If one system exerts a force on a second system, then the second system exerts a reaction force on the first system, whether or not the systems are accelerating. These forces of action and reaction between the systems are equal in magnitude, opposite in direction and co-linear.

3.2 Newton's laws applied to translational motion

Newton provides a link, through the second law, between the acceleration experienced by a system and the resultant or unbalanced force acting on it. We can write the second law mathematically as follows:

$$a \propto \frac{F}{m}$$

$$a = k \cdot \frac{F}{m}$$

If we choose units to make the constant $k = 1$ and rearrange the equation in terms of F then:

$$F = ma \tag{3.1}$$

A suitable set of units is provided by the SI system (Système International d'Unités) where:

F = the total force (N) acting on the system
m = mass (kg) of the system
a = acceleration ($\mathrm{m\,s^{-2}}$) of the system.

It follows from setting $k = 1$ that $1\,\mathrm{N} \equiv 1\,\mathrm{kg\,m\,s^{-2}}$. Mass is a scalar while force and acceleration are vectors. Hence the equation developed from Newton's second law can be written in component form. When applying the equations in a problem, it is important to indicate the axis system to be used.

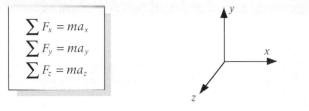

$$\sum F_x = ma_x$$
$$\sum F_y = ma_y$$
$$\sum F_z = ma_z$$

(3.2)

Newton's second law proves a powerful tool for solving problems involving individual objects or complete systems in translation. We can use the second law to determine either the acceleration of the object, if all the applied forces are known, or the resultant force acting on the object, if the acceleration is known. Note that if the applied forces are constant, so also must be the acceleration, and the kinematic formulae developed in Chapter 2 will apply.

To solve a translational dynamics problem draw a 'free-body diagram' of the object, isolating it from its surroundings. A free-body diagram is a diagram showing the forces acting *on* the object. Remember to include weight as a force acting on the object.

The problems are best solved in a systematic fashion as suggested in the following four-step procedure:

1. Draw a sketch of the problem.
2. Draw a free-body diagram of the object.
   To draw the free-body diagram first sketch the outline of the object, isolating it from its surroundings. Show on the outline all the forces acting on the object, including its weight. This can be done efficiently by travelling around the outline and attaching force arrows to the diagram where each applied force, supporting surface or fixing is encountered.
   (a) Supporting surfaces should be represented by two force arrows on the free-body diagram. These forces are: a 'normal reaction', R, which is perpendicular to the surface acting towards the object and a 'friction force', S, which is parallel to the surface and acting in a direction opposing the motion of the object. W represents the object's weight.

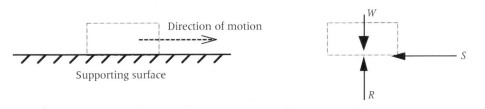

The friction force is proportional to the normal reaction force and a coefficient of friction term, μ, which is dependent upon the relative roughness of the surface and the object. Where sliding is taking place S, R and μ are related as follows:

$$S = \mu R$$

(3.3)

Note that this equation only holds good where sliding is actually occurring and that generally $S \leq \mu R$.

(b) With structural members fixed to the ground, as we shall encounter later in this chapter, there are three types of fixings: full fixing, pin fixing and roller fixing. The conventions for showing these and the equivalent force representations are shown below.

Full fixing   *Pin fixing*   *Roller fixing*

Full fixing restrains the member against both horizontal and vertical translation. Additionally the member end is prevented from rotating. Hence the force representation consists of a vertical force, a horizontal force and a fixing moment. A pin fixing restrains the member against horizontal and vertical translation but the member end is free to rotate. The corresponding representation consists of a horizontal and a vertical force. A roller fixing restrains the member against vertical translation but permits horizontal translation and rotation of the member end. The representation consists of a vertical force.

3. On the free-body diagram show an x–y axes system. Use Newton's second law to generate equations relating force to mass and acceleration. Apply the law independently to the translational motions in the x and y directions, taking care to define positive and negative values strictly in accordance with the adopted axes system.

4. Solve the resulting simultaneous equations to determine the unknown applied force or resulting acceleration.

EXAMPLE 3.1

A crate with a mass of 60 kg is pulled along a horizontal surface by a horizontal force of 500 N. The coefficient of friction, μ, between the crate and the surface is 0.3. Determine the acceleration of the crate.

Solution

First draw a sketch of the problem, then a free-body diagram of the crate:

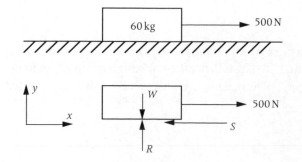

Apply Newton's second law in the direction of the y-axis and note that there is no movement of the system in this direction and hence that a_y must be zero:

$$\sum F_y = ma_y$$
$$= 0$$
$$R - W = 0$$
$$R = W$$
$$= (60 \times 9.81)$$
$$= 588.6 \, \text{N}$$

The crate is slipping along the plane. Therefore the friction force, S, can be calculated:

$$S = \mu R$$
$$= 0.3 \times 588.6$$
$$= 176.58 \, \text{N}$$

Apply Newton's second law in the direction of the x-axis:

$$\sum F_x = ma_x$$
$$500 - S = ma_x$$
$$500 - 176.58 = 60a_x$$
$$a_x = 5.39 \, \text{m s}^{-2}$$

Hence the acceleration of the crate is $5.39 \, \text{m s}^{-2}$. The implied positive answer tells us that the acceleration is to the right in the direction of the positive x-axis.

EXAMPLE 3.2

The crate described in Example 3.1 is now pulled along the same surface with the 500 N force inclined at 10° to the horizontal. Calculate the new acceleration of the crate.

Solution

Resolve the 500 N force into its components aligned with the x and y axes.

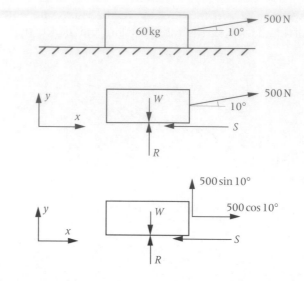

Apply Newton's second law to the forces aligned along the y-axis:

$$\sum F_y = ma_y = 0$$

$$R + 500 \sin 10° - W = 0$$

$$R = W - 500 \sin 10°$$

$$= (60 \times 9.81) - (500 \sin 10°)$$

$$= 501.776 \text{ N}$$

The crate is slipping therefore we can calculate the friction force S:

$$S = \mu R$$

$$= 0.3 \times 501.776$$

$$= 150.533 \text{ N}$$

Apply Newton's second law to the forces which are aligned in the direction of the x-axis:

$$\sum F_x = ma_x$$

$$500 \cos 10° - S = 60a_x$$

$$500 \cos 10° - 150.533 = 60a_x$$

$$a_x = 5.70 \text{ m s}^{-2}$$

The acceleration of the crate is now 5.70 m s^{-2}, horizontally to the right.

EXAMPLE 3.3

A crate of mass 60 kg is pulled up a plane which is inclined at an angle of 20° to the horizontal. The coefficient of friction between the crate and the surface is 0.3. Determine the pulling force P, parallel to the surface, which will accelerate the crate up the slope at 3 m s^{-2}.

Solution

Align the x and y axes parallel and perpendicular to the plane. Resolve the weight force into its components in the x and y axes.

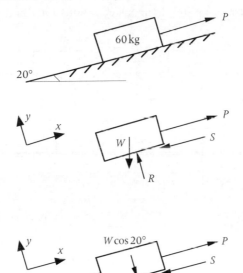

Apply Newton's second law to the forces aligned along the y-axis:

$$\sum F_y = ma_y = 0$$

$$R - W \cos 20° = 0$$

$$R = (60 \times 9.81) \cos 20°$$

$$= 553.1 \text{ N}$$

Apply Newton's second law to the forces which are aligned in the direction of the x-axis:

$$\sum F_x = ma_x$$

$$R - W \sin 20° - S = ma_x$$

$$P = (60 \times 9.81) \sin 20° + (0.3 \times 553.1) + (60 \times 3)$$

$$= 547.24 \text{ N}$$

The required force has a magnitude of 547.24 N.

EXAMPLE 3.4

Block 1, of mass 2 kg, is connected by a wire to block 2. Block 2 has a mass of 3 kg. The two blocks are pulled along a horizontal surface by a horizontal force of 18 N applied to block 1. The coefficient of friction between the surface and the blocks is 0.2. Calculate the acceleration of the blocks and the tension force in the connecting wire.

Solution

Sketch separate free-body diagrams of each block. Apply Newton's second law to each diagram to generate equations in T, the tension in the wire, and a_x, the acceleration. Solve the resulting simultaneous equations for T and a_x.

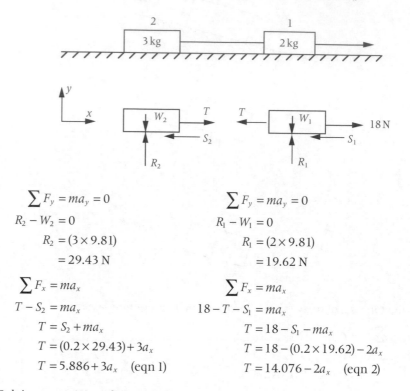

$$\sum F_y = ma_y = 0$$
$$R_2 - W_2 = 0$$
$$R_2 = (3 \times 9.81)$$
$$= 29.43\,\text{N}$$

$$\sum F_x = ma_x$$
$$T - S_2 = ma_x$$
$$T = S_2 + ma_x$$
$$T = (0.2 \times 29.43) + 3a_x$$
$$T = 5.886 + 3a_x \quad (\text{eqn 1})$$

$$\sum F_y = ma_y = 0$$
$$R_1 - W_1 = 0$$
$$R_1 = (2 \times 9.81)$$
$$= 19.62\,\text{N}$$

$$\sum F_x = ma_x$$
$$18 - T - S_1 = ma_x$$
$$T = 18 - S_1 - ma_x$$
$$T = 18 - (0.2 \times 19.62) - 2a_x$$
$$T = 14.076 - 2a_x \quad (\text{eqn 2})$$

Solving eqns (1) and (2) simultaneously:

$$5.886 + 3a_x = 14.076 - 2a_x$$
$$a_x = \frac{14.076 - 5.886}{5}$$
$$a_x = 1.638\,\text{m s}^{-2}$$

Substituting $1.638\,\text{m s}^{-2}$ for a_x in eqn (2):

$$T = 14.076 - 2(1.638)$$
$$T = 10.8\,\text{N}$$

The acceleration of the system of crates is $1.638\,\text{m s}^{-2}$ horizontally to the right and the tension force in the connecting wire is $10.8\,\text{N}$.

EXAMPLE 3.5

A 2 kg block is pulled along a horizontal surface by a wire which passes over a small pulley wheel and is connected to a second block of mass 3 kg hanging vertically. The coefficient of friction between the 2 kg block and the surface is 0.2. Determine the acceleration of the system and the force in the connecting wire. The effect of the small pulley wheel can be neglected.

Solution

Generate simultaneous equations by analysing each block separately and then solve to determine T and a_x. A consistent system of axes is required for the free-body diagrams. Choose x as the direction of motion in both cases.

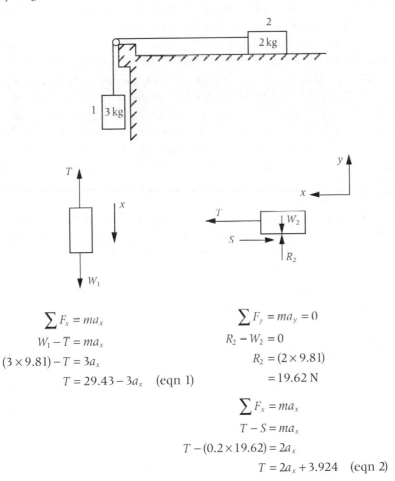

$$\sum F_x = ma_x$$
$$W_1 - T = ma_x$$
$$(3 \times 9.81) - T = 3a_x$$
$$T = 29.43 - 3a_x \quad \text{(eqn 1)}$$

$$\sum F_y = ma_y = 0$$
$$R_2 - W_2 = 0$$
$$R_2 = (2 \times 9.81)$$
$$= 19.62 \, \text{N}$$

$$\sum F_x = ma_x$$
$$T - S = ma_x$$
$$T - (0.2 \times 19.62) = 2a_x$$
$$T = 2a_x + 3.924 \quad \text{(eqn 2)}$$

Solving eqns (1) and (2) simultaneously:

$$29.43 - 3a_x = 2a_x + 3.924$$
$$5a_x = 25.506$$
$$a_x = 5.101 \, \text{m s}^{-2}$$

Substituting back into eqn (1):

$$T = 29.43 - (3 \times 5.101)$$
$$T = 14.126 \, \text{N}$$

The acceleration of the system is $5.101\,\mathrm{m\,s^{-2}}$ in the positive x-direction and the tension force in the connecting wire is $14.126\,\mathrm{N}$.

3.3   Further discussion on Examples 3.1 to 3.5

The examples assume that the motion experienced in each case is purely translational. We can only be sure that this is so if the forces acting on the systems, in this case the crates and blocks, are concurrent. If the forces are not concurrent then the possibility of rotation arises. The forces in the examples are not concurrent. The pulling forces and the friction forces, for example, are parallel. This doesn't mean, necessarily, that rotations will take place. It is all a matter of whether or not the rotational effects of the forces, or moments, balance out. Moments are dealt with in Chapter 4 but it is sufficient to appreciate here that we are assuming that the relative horizontal and vertical dimensions of the crates and blocks are such as to avoid any possibility of them tipping over on their front edges when the pulling forces are applied.

We chose to ignore the effect of the small pulley in Example 3.5. This is reasonable at this stage. However it is a moving part of the system; it is rotating as the wire passes around the rim. For large pulleys or axles, this rotation is significant and affects, among other things, the tension forces in the wire on either side of the pulley. Examples of systems in which the effects of the pulley rotation is taken into account are dealt with in Chapter 4.

3.4   Problems

In questions **3.1** to **3.4** the block has a mass of 45 kg and the coefficient of friction between the block and the plane is 0.25.

3.1   Determine the force P required to cause an acceleration of $4\,\mathrm{m\,s^{-2}}$.

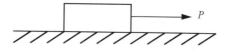

3.2   Calculate P, such that the acceleration up the slope is $2\,\mathrm{m\,s^{-2}}$.

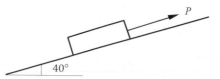

3.3 Calculate the acceleration of the block.

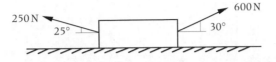

3.4 Determine the force P required to restrict the block such that it slips down the slope with an acceleration of $1.5\,\mathrm{m\,s^{-2}}$.

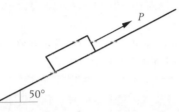

In questions **3.5** and **3.6**, determine the acceleration of the systems and the tension forces in the connecting wires.

3.5

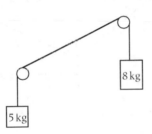

3.6 The coefficient of friction between the block and the plane is 0.3.

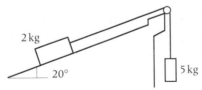

3.5 | Translational equilibrium

Many solid mechanics systems do not move. They neither translate nor rotate; they are static. Systems of this type are said to be in equilibrium. Civil engineering structures such as building frames, walls and bridges are in equilibrium and Newton's laws can be used to determine the forces of action and reaction on these systems.

For translational equilibrium it follows that the acceleration term, a_x, in the second law equation must be zero and hence for static problems we have:

$$\sum F_x = 0$$
$$\sum F_y = 0$$
$$\sum F_z = 0$$

(3.4)

EXAMPLE 3.6

A 50 kN weight is suspended from two wires P and Q attached to a roof beam as shown. Determine the forces in the two wires.

Solution

The wires meet at joint 1. Draw a free-body diagram showing the forces acting on this joint. Resolve the forces to align them with the *x* and *y* axes.

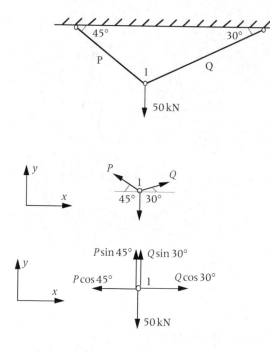

The problem is quite clearly a static one. The system is in equilibrium. Generate two equations in *P* and *Q* from considering equilibrium of the forces in the *x*-direction and in the *y*-direction.

$$\sum F_x = 0$$

$$Q \cos 30° - P \cos 45° = 0$$

$$0.866Q - 0.707P = 0 \quad \text{(eqn 1)}$$

$$\sum F_y - 0$$

$$Q \sin 30° + P \sin 45° - 50 = 0$$

$$0.5Q + 0.707P = 50 \quad \text{(eqn 2)}$$

Solve eqns (1) and (2) simultaneously. Add the equations to eliminate P.

$$1.366Q = 50$$

$$Q = 36.603 \, \text{kN}$$

Note that the value obtained for Q is positive. This indicates that the direction of the arrow of the Q force on the free-body diagram was correct. It is pulling on the joint and Q is in tension. If a negative solution had occurred, then this would have indicated that the assumed direction of the arrow was incorrect and that Q was in fact pushing on the joint and was in compression.

Now substitute $36.603 \, \text{kN}$ for Q in eqn (2) to determine the value of P.

$$(0.5)(36.603) + 0.707P = 50$$

$$P = \frac{50 - (0.5)(36.603)}{0.707}$$

$$P = 44.835 \, \text{kN}$$

Again a positive answer is obtained, so the assumed direction for P is also correct. P is in tension.

EXAMPLE 3.7

Determine the forces in the two members, P and Q, of the pin-jointed structure shown.

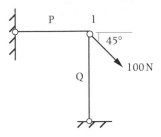

Solution

Draw the free-body diagram of the joint where P, Q and the 100 N force meet. The 'correct' direction for the arrows showing the forces acting on the joint can be determined as follows. To find the direction of the force P on the free-body diagram, consider what would happen to the structure if this member were removed. The structure would move to the right. The member which we are imagining to be removed must be preventing this from happening; it must be pulling on the joint to the left. A similar thought experiment applied to member Q shows it to be pushing upwards on the joint.

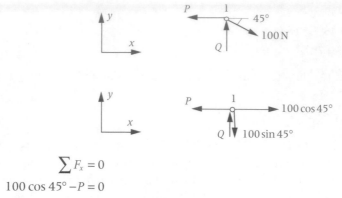

$$\sum F_x = 0$$
$$100 \cos 45° - P = 0$$
$$P = 70.71\,\text{N}$$

The positive answer obtained confirms the assumption that P is pulling on the joint. P is a tie member subjected to a tension force of 70.71 N.

$$\sum F_y = 0$$
$$Q - 100 \sin 45° = 0$$
$$Q = 70.71\,\text{N}$$

The positive solution confirms the assumption that Q is pushing upwards on the joint. Q is therefore a strut subjected to a compressive force of 70.71 N.

EXAMPLE 3.8

Determine the forces in members P and Q.

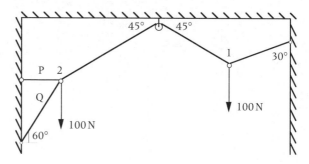

Solution

Since we can only generate two equations, one each from the x and y directions, it is not possible to directly solve forces where there are more than two unknowns such as at joint 2. The solution is to first consider joint 1 to determine the forces in the two wires and then to move on to joint 2.

First consider translational equilibrium at joint 1. Denote the unknown forces at the joint by R and S. The direction of both of these forces must be pulling on joint 1.

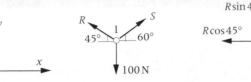 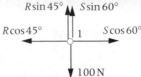

$$\sum F_x = 0$$

$$S\cos 60° - R\cos 45° = 0$$

$$0.5S - 0.707R = 0 \quad (\text{eqn 1})$$

$$\sum F_y = 0$$

$$S\sin 60° + R\sin 45° - 100 = 0$$

$$0.866S + 0.707R = 100 \quad (\text{eqn 2})$$

Solve eqns (1) and (2) simultaneously. Add the equations to eliminate R and then back-substitute the value found for S into eqn (2) to determine the value of R.

$$1.366S = 100$$

$$S = \frac{100}{1.366}$$

$$S = 73.206 \text{ N}$$

$$(0.5)(73.206) - 0.707R = 0$$

$$R = \frac{(0.5)(73.206)}{0.707}$$

$$R = 51.773 \text{ N}$$

The force in the left-hand wire at joint 1 is therefore 51.773 N. This wire is connected via the small pulley to joint 2. The force applied by this wire at joint 2 must also be 51.773 N and again for equilibrium it must be pulling on joint 2. Note also that if we consider the direction of the forces P and Q to prevent movement at joint 2, then P pulls on the joint and Q pushes. Now consider translational equilibrium at joint 2.

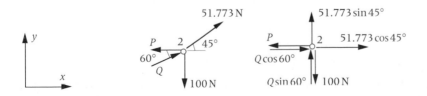

It is preferable in this instance to consider vertical equilibrium first as the resulting equation immediately generates the solution for Q independently of P.

$$\sum F_y = 0$$

$$51.773 \sin 45° + Q \sin 60° - 100 = 0$$

$$Q = \frac{100 - 51.773 \sin 45°}{\sin 60°}$$

$$Q = 73.198 \text{ N}$$

Horizontal equilibrium will now provide the solution for P:

$$\sum F_x = 0$$

$$Q \cos 60° + 51.773 \cos 45° - P = 0$$

$$P = 51.773 \cos 45° + 73.198 \cos 60°$$

$$P = 73.208 \text{ N}$$

We have determined therefore that Q is a strut supporting a compressive force of 73.198 N and P is a tie carrying a tensile force of 73.208 N.

3.6 | Method of joints

As seen in Example 3.8, pin-jointed structures with several joints can be solved by systematically working from joint to joint provided that there are no more than two unknowns at any stage in the process. Not surprisingly, this solution technique is known as the *method of joints*.

EXAMPLE 3.9

Determine the forces in each member of the frame shown below.

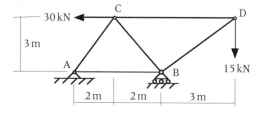

Solution

Note the difference between the detail of the reaction at A and that at B. At A, both horizontal and vertical forces can be resisted. B is in the form of a roller and does not offer any resistance horizontally. At B there is purely a vertical reaction. First draw the free-body diagram showing the reactions at A and B acting in the positive x and y directions.

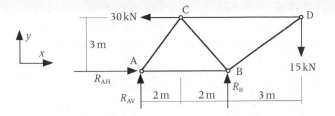

Start at joint D, where there are only two unknowns, the forces in members CD and BD. Using the technique explained in Example 3.7 it is easily seen that BD must be a strut and CD a tie member.

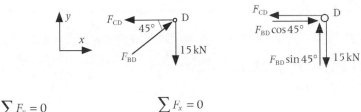

$$\sum F_y = 0$$
$$F_{BD} \sin 45° - 15 = 0$$
$$F_{BD} = \frac{15}{\sin 45°}$$
$$F_{BD} = 21.213 \text{ kN} \quad \text{strut}$$

$$\sum F_x = 0$$
$$F_{BD} \cos 45° - F_{CD} = 0$$
$$F_{CD} = F_{BD} \cos 45°$$
$$F_{CD} = 21.213 \cos 45°$$
$$F_{CD} = 15 \text{ kN} \quad \text{tie}$$

Now move on to joint C where the remaining unknowns are the forces in members CB and AC. Member AC is definitely a strut but CB is less clear. Let's assume it is a strut. If we are wrong then the sign of the answer will inform us of the error.

$$\sum F_x = 0$$
$$15 + F_{AC} \cos 56.31° - 30 - F_{CB} \cos 56.31° = 0$$
$$0.555 F_{AC} - 0.555 F_{CB} = 15 \quad \text{(eqn 1)}$$

$$\sum F_y = 0$$
$$F_{AC} \sin 56.31° + F_{CB} \sin 56.31° = 0$$
$$F_{AC} = -F_{CB} \quad \text{(eqn 2)}$$

From eqn (2) the negative sign makes it clear that if AC is a strut then CB must be a tie. There is no need to amend the free-body diagram at this stage; it is best just to continue and to sort things out at the end. Substitute for CB in eqn (1):

$$0.555 F_{AC} - 0.555(-F_{AC}) = 15$$
$$F_{AC} = \frac{15}{2(0.555)}$$
$$= 13.514 \text{ kN} \quad \text{strut}$$

Substituting into eqn (2) gives $F_{CB} = -13.514\,\text{kN}$. The negative sign tells us that our original assumption that this member was a strut was wrong and in fact it is a tie carrying a force of $13.514\,\text{kN}$. Now move on to joint B.

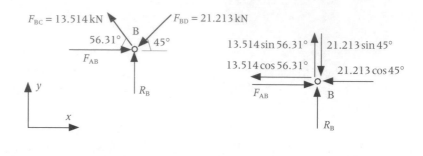

$$\sum F_x = 0$$

$$F_{AB} - 13.514\cos 56.31° - 21.213\cos 45° = 0$$

$$F_{AB} = 22.496\ \text{kN}\quad \text{strut}$$

$$\sum F_y = 0$$

$$R_B + 13.514\sin 56.31° - 21.213\sin 45° = 0$$

$$R_B = 3.756\ \text{kN}$$

Finally we can determine the horizontal and vertical components of the reaction at joint A.

$$\sum F_x = 0$$

$$R_{AH} - 13.514\cos 56.31° - 22.496 = 0$$

$$R_{AH} = 29.992\ \text{kN}$$

$$\sum F_y = 0$$

$$R_{AV} - 13.514\sin 56.31° = 0$$

$$R_{AV} = 11.244\ \text{kN}$$

It is a good check to consider the overall equilibrium of the frame.

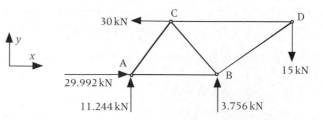

$$\sum F_x = 29.992 - 30 \qquad \sum F_y = 11.244 + 3.756 - 15$$
$$= -0.008 \qquad\qquad = 0$$

The slight error is due to rounding. An alternative approach to solving pin-jointed frames using moment equilibrium, the *method of sections*, is shown in Chapter 4.

3.7 Further discussion on Examples 3.6 to 3.9

Examples 3.6 to 3.9 are of pin-jointed frames. In this type of frame each connection can be considered as consisting of a single pin. Such connections allow the ends of the members to rotate individually under loading. Moreover the loads are applied only at the joints and not along the lengths of the frame members. The members of pin-jointed frames of this type carry only axial loads and are not subjected to any bending effects. Rigid joints would transfer any rotation from one member into the other members surrounding the joint and are not considered in this text.

It is important to remember that it is the free-body diagrams of the pin joints themselves that are being drawn. Thus, as in all free-body diagrams, we show the forces acting on the free-body. In other words we see on these drawings what the connecting members are doing to the joint, pushing or pulling. If we were to draw the free-body diagrams of the members then we would be showing the forces acting on the members from the pins. Numerically the force values would be the same but the direction arrows would be reversed. In the illustration below, member P is in tension and is a tie, while member Q is in compression and is a strut.

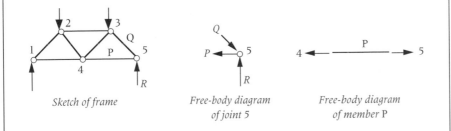

Sketch of frame   *Free-body diagram of joint 5*   *Free-body diagram of member* P

In Example 3.9 we were fortunate to have a joint, D, at which there were only two unknowns and where we could begin our analysis of the frame. This is not always the case. Compare the two frames shown overleaf; the upper frame is Example 3.9.

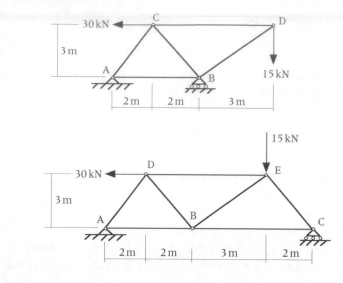

In the lower frame there is no equivalent starting point. However, the fixing at C represents a single vertical force (the reaction force) at C. If we could determine this reaction force then there would only remain two unknowns at C, the force in member BC and the force in member EC. We could then begin the method of joints procedure at C. The calculation of reaction forces is explained in Chapter 4, Section 4.5.

3.8 | Problems (continued)

For each of the pin-jointed frames in questions **3.7** to **3.10**, determine the forces in members P and Q.

3.7

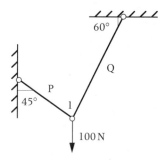

3.8

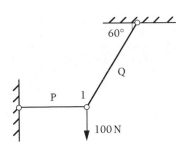

3.9

3.10

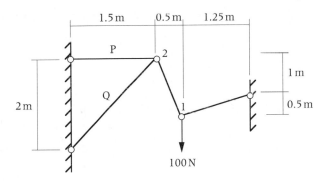

3.11   Using the method of joints, determine the forces in each member of the pin-jointed frame below.

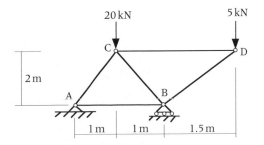

CHAPTER 4

Newton's Laws of Motion: Rotational Motion

In Chapter 3 we considered the solution of problems where the motion of the system was purely translational. In many systems the actions of the forces additionally give rise to rotations. This is particularly true of systems where the applied forces are non-concurrent. In this chapter we will look firstly at systems which are purely rotational and then consider systems in which rotation and translation are both present. Finally, systems in rotational equilibrium are examined leading to the determination of reaction forces in statics problems.

The following concepts are introduced in this chapter:

- *Moments*
- *Torques*
- *Second moment of mass*
- *Rotational acceleration*
- *Radius of gyration*
- *Rotational equilibrium*
- *Reactions*
- *Method of sections*

4.1 | Newton's second law in rotational form

Newton's second law can be applied to rotational motion. For rotational motion Newton's second law takes the following form:

$$\sum M = I\alpha$$

(4.1)

where M = moments and torques (Nm) acting on the system about the mass centre

I = second moment of mass (kgm^2) of the system

α = rotational acceleration (rads^{-2}) about the mass centre.

The solution strategy for rotational problems follows a similar pattern to that for translational problems.

1. Draw a sketch of the problem.

2. Draw a free-body diagram of an appropriate element of the system showing all the forces and applied torques acting on it, including its weight if this is significant.

3. First calculate the second moment of mass (common shapes such as the solid cylinder, the thin wheel and the rectangular bar are shown in Table 1.1 on page 25), then use Newton's second law in rotational form to generate an equation relating moments to second moment of mass and rotational acceleration.

4. In order to generate sufficient equations to solve the problem you will generally also need to consider the translational characteristics of the situation and to use the translational form of the second law.

5. Use the link between the instantaneous tangential acceleration, a_{tan}, and the rotational acceleration, α, to set all the equations in terms of either a_{tan} or α.

$$a_{tan} = \alpha r$$

(4.2)

6. Solve the resulting equations to determine the unknown applied forces or resulting accelerations.

EXAMPLE 4.1

This example is typical of fixed axes rotational problems. The winding drum shown rotates about a fixed axle at A. An anticlockwise torque, T, of 700 Nm is applied to the drum via the axle. A wire is wound round the circumference of the drum and a force, P, of 4000 N is applied to the end of the wire. Determine the reaction at the axle and the angular acceleration of the drum.

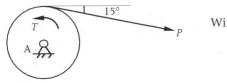

Winding drum data:

diameter = 0.5 m
length   = 0.4 m
density = 8000 kg m^{-3}

Solution

Draw a free-body diagram of the winding drum. The winding drum is in fixed axis rotation so first determine its second moment of mass.

$$m = \rho V$$
$$= 8000(\pi \times 0.25^2)(0.4)$$
$$= 628 \text{ kg}$$
$$I = \frac{mr^2}{2}$$
$$= \frac{628 \times 0.25^2}{2}$$
$$= 19.625 \text{ kg m}^2$$

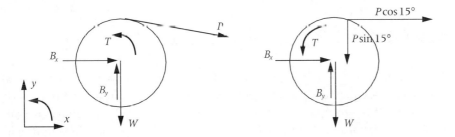

Apply the translational form of Newton's second law in first the x and then the y directions to determine B_x and B_y. Note that there is no movement in either the x or y directions and that these two equations therefore take their equilibrium form.

$$\sum F_x = 0$$
$$B_x + P \cos 15° = 0$$
$$B_x = -4000 \cos 15°$$
$$B_x = -3864 \text{ N}$$
$$\sum F_y = 0$$
$$B_y - W - P \sin 15° = 0$$
$$B_y = (628 \times 9.81) + (4000 \sin 15°)$$
$$B_y = 7197 \text{ N}$$

Determine the resultant, B, of the two components of the reaction force at the axle, B_x and B_y.

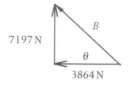

7197 N

3864 N

$$B = \sqrt{3864^2 + 7197^2}$$

$$B = 8169 \text{ N}$$

$$\theta = \tan^{-1}\left(\frac{7197}{3864}\right)$$

$$\theta = 61.8°$$

The reaction at the axle is 8169 N at 61.8° ⬊.

Apply the rotational form of Newton's second law to the free-body diagram of the winding drum, taking moments about the mass centre A.

$$\sum M = I\alpha$$

$$T - Pr = I\alpha$$

$$\alpha = \frac{T - Pr}{I}$$

$$\alpha = \frac{700 - (4000 \times 0.25)}{19.625}$$

$$\alpha = -15.287 \text{ rad s}^{-2}$$

The rotational acceleration of the winding drum is 15.287 rad s^{-2}. The negative sign indicates that the rotational acceleration is clockwise.

EXAMPLE 4.2

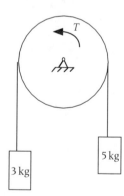

The solution of the following problem involves the use of both the translational and rotational equations. Independent free-body diagrams of the two translating masses and the rotating drum are constructed and analysed separately, generating three equations in three unknowns which can then be solved. Note in particular that the tension forces in the section of wire attached to the left- and right-hand masses are not equal.

A drum with diameter of 1.5 m, length 0.3 m and density 10 500 kg m^{-3} supports two masses as shown. The drum is being driven anticlockwise by a torque T, of 250 Nm. Determine the rotational acceleration of the drum, the tensions in the support cable and the translational acceleration of the two masses.

Solution

First draw the free-body (FBD) diagrams of each of the masses and from these construct two equations involving T_1, T_2 and a_x. Note that we will assume that the resulting motion of the system is such that the drum rotates anticlockwise. It is essential that consistency is maintained throughout the problem in terms of the set direction for positive motion and thus the x-axis will be positive upwards for the 5 kg mass and positive downwards for the 3 kg mass.

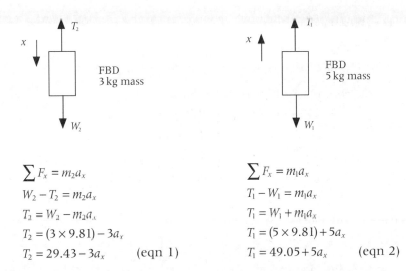

$$\sum F_x = m_2 a_x$$
$$W_2 - T_2 = m_2 a_x$$
$$T_2 = W_2 - m_2 a_x$$
$$T_2 = (3 \times 9.81) - 3a_x$$
$$T_2 = 29.43 - 3a_x \qquad \text{(eqn 1)}$$

$$\sum F_x = m_1 a_x$$
$$T_1 - W_1 = m_1 a_x$$
$$T_1 = W_1 + m_1 a_x$$
$$T_1 = (5 \times 9.81) + 5a_x$$
$$T_1 = 49.05 + 5a_x \qquad \text{(eqn 2)}$$

Next draw the free-body diagram of the drum and develop from the rotational equation of motion the third equation which relates T_1, T_2 and α.

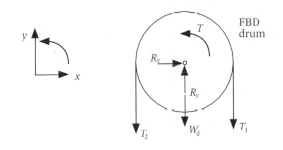

$$I = \frac{mr^2}{2}$$
$$= 0.5(10\,500 \times \pi \times 0.75^2 \times 0.3)(0.75^2)$$
$$= 1566 \text{ kg m}^2$$

$$\sum M = I\alpha$$
$$(T_2 \times r) - (T_1 \times r) + T = 1566\alpha$$
$$0.75(T_2 - T_1) + 250 = 1566\alpha \qquad \text{(eqn 3)}$$

Hence we now have the following three equations with four unknowns, T_1, T_2, a_x and α:

$$T_1 = 49.05 + 5a_x \quad \text{(eqn 1)}$$
$$T_2 = 29.43 - 3a_x \quad \text{(eqn 2)}$$
$$0.75(T_2 - T_1) + 250 = 1566\alpha \quad \text{(eqn 3)}$$

Since $a_{\text{tan}} = \alpha r$ we can replace a_x in eqns (1) and (2) by αr. Hence:

$$T_1 = 49.05 + (5)(\alpha)(0.75)$$
$$T_1 = 49.05 + 3.75\alpha \qquad \text{(eqn 4)}$$

$$T_2 = 29.43 - (3)(\alpha)(0.75)$$
$$T_2 = 29.43 - 2.25\alpha \qquad \text{(eqn 5)}$$

Substituting for T_1 and T_2 from eqn (4) and eqn (5) in eqn (3) gives:

$$0.75(29.43 - 2.25\alpha - 49.05 - 3.75\alpha) + 250 = 1566\alpha$$
$$250 - 14.715 - 4.5\alpha = 1566\alpha$$
$$\alpha = 0.150\,\text{rad}\,\text{s}^{-1}$$

The answer is positive and therefore the rotational acceleration is anticlockwise.

If we now substitute the value obtained for α into eqns (4) and (5) then the wire tensions T_1 and T_2 can be obtained:

$$T_1 = 49.05 + (3.75)(0.150)$$
$$T_1 = 49.612\,\text{N}$$

$$T_2 = 29.43 - (2.25)(0.150)$$
$$T_2 = 29.093\,\text{N}$$

The tensions in the wires supporting the 5 kg and 3 kg masses are 49.612 N and 29.03 N respectively.

By substituting the value obtained for T_1 into eqn (1) we can determine the translational accelerations of the two masses:

$$49.612 = 49.05 + 5a_x$$
$$a_x = 0.112\,\text{m}\,\text{s}^{-2}$$

According to the axes system set up at the outset of the solution, the positive value obtained for a_x indicates that the 5 kg mass is accelerating upwards at $0.112\,\text{m}\,\text{s}^{-2}$ and the 3 kg mass downwards at $0.112\,\text{m}\,\text{s}^{-2}$.

EXAMPLE 4.3

The solution of the next two problems requires the use of both the rotational and translational dynamics sets of equations applied to a single object. Determine the acceleration of the centre of the wheel shown. The wheel has a mass of 65 kg and a radius of gyration of 0.22 m. The outside wheel diameter is 0.6 m and the axle diameter is 0.4 m. It can be assumed that no slipping takes place.

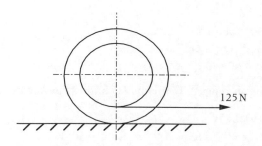

Solution

As usual, start by drawing the free-body diagram of the wheel.

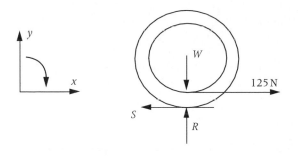

Firstly consider the translational motion equations:

$$\sum F_x = ma_x$$
$$125 - S = 65a_x$$
$$S = 125 - 65a_x \quad \text{(eqn 1)}$$

$$\sum F_y = 0$$
$$R - W = 0$$
$$R = W$$
$$R = 65 \times 9.81$$
$$R = 637.65 \text{ N}$$

It is important to note here that, since the wheel is rolling, then $S \le \mu R$ and it is therefore not possible at this stage to solve eqn (1) directly. The equation $S = \mu R$ only applies when surfaces are sliding over each other when the friction force is fully developed. This would only be the case in this example if the wheel was fully locked.

Now consider the rotational motion equation:

$$\sum M = I\alpha$$
$$(S \times 0.3) - (125 \times 0.2) = I\alpha$$

For a wheel $I = mk^2$, and hence:

$$(S \times 0.3) - (125 \times 0.2) = 65 \times 0.22^2 \alpha$$
$$\alpha = \frac{0.3S - 25}{3.146} \quad \text{(eqn 2)}$$

In eqn (1) the unknowns are S and a_x while in eqn (2) the unknowns are S and α. However, provided that the wheel is rolling without slipping, $a_x = r\alpha$ and eqn (2) can be rewritten as:

$$\frac{a_x}{r} = \frac{0.3S - 25}{3.146}$$

$$a_x = \frac{(0.3S - 25)(0.3)}{3.146}$$

$$0.09S - 7.5 = 3.146a_x$$

$$S = 83.333 + 34.956a_x \quad \text{(eqn 3)}$$

Equations (1) and (3) can be solved simultaneously to determine the translational acceleration of the centre of the wheel:

$$125 - 65a_x = 83.333 + 34.956a_x$$

$$99.956a_x = 41.667$$

$$a_x = 0.417 \, \text{m s}^{-2}$$

It follows that the rotational acceleration of the centre of the wheel is given by:

$$\alpha = \frac{a_x}{r}$$

$$\alpha = \frac{0.417}{0.3}$$

$$\alpha = 1.39 \, \text{rad s}^{-2}$$

Notice also that from eqn (1) it is possible to determine the friction force, S, that is generated as the wheel rotates:

$$S = 125 - 65a_x$$

$$S = 125 - (65 \times 0.417)$$

$$S = 97.895 \, \text{N}$$

Thus the minimum coefficient of friction required to ensure that the wheel rotates rather than slides is given by:

$$\mu_{\text{min}} = \frac{S}{R} = \frac{97.895}{637.65} = 0.15$$

EXAMPLE 4.4

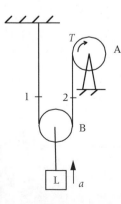

Calculate the tension in the cable at points 1 and 2 and determine the torque required to be applied to the winding drum, A, to cause the load, L, to be raised with an acceleration of $3.5 \, \text{m s}^{-2}$.

Data:
winding drum, A
      diameter = 0.5 m
      2nd moment of mass = $0.8 \, \text{kg m}^2$
lifting pulley, B
      diameter = 0.4 m
      2nd moment of mass = $0.2 \, \text{kg m}^2$ load, L
      mass = 20 kg   mass = 150 kg

Solution

First draw a free-body diagram of lifting pulley B, the load L and the cable between points 1 and 2.

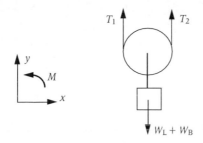

Apply Newton's second law in the y-direction:

$$\sum F_y = ma_y$$

$$T_1 + T_2 - (W_L + W_B) = (m_L + m_B)a_y$$

$$T_1 + T_2 = (W_L + W_B) + (m_L + m_B)a_y$$

$$T_1 + T_2 = (150 \times 9.81 + 20 \times 9.81) + (150 + 20)3.5$$

$$T_1 + T_2 = 2262.7 \qquad \text{(eqn 1)}$$

Consider rotation:

$$\sum M_B = I_B \alpha_B$$

$$(T_2 \times 0.2) - (T_1 \times 0.2) = 0.2\alpha$$

$$T_2 - T_1 = \alpha_B$$

$$a_y = r_B \alpha_B$$

$$\therefore T_2 - T_1 = \frac{a_y}{r_B}$$

$$T_2 - T_1 = \frac{3.5}{0.2}$$

$$T_2 - T_1 = 17.5 \qquad \text{(eqn 2)}$$

Solve for T_1 and T_2 by adding eqns (1) and (2):

$$2T_2 = 2280.2$$

$$T_2 = 1140.1\,\text{N}$$

$$T_1 = 2262.7 - 1140.1$$

$$T_1 = 1122.6\,\text{N}$$

The tensions in the cable at points 1 and 2 respectively are 1122.6 N and 1140.1 N. Now draw a free-body diagram of the winding drum, A, and the attached cable.

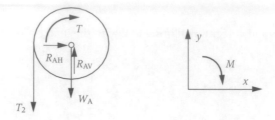

Consider rotation:

$$\sum M_A = I_A \alpha_A$$
$$T - (0.25T_2) = 0.8\alpha_A$$
$$T = 0.25T_2 + 0.8\alpha_A \quad \text{(eqn 3)}$$

Use the link equation and note that the tangential acceleration of the winding wheel, A, is twice the vertical acceleration of the load, L.

$$a_y = r_A \alpha_A$$
$$(2 \times 3.5) = 0.25\alpha_A$$
$$\alpha_A = 28\ \text{rad}\,\text{s}^{-2}$$

Substituting $28\ \text{rad}\,\text{s}^{-2}$ for the rotational acceleration of the winding drum, A, into eqn (3):

$$T = (0.25 \times 1140.1) + (0.8 \times 28)$$
$$T = 307.425\ \text{N}\,\text{m}$$

The required torque to be applied to the winding drum, A, is 307.425 N m clockwise.

EXAMPLE 4.5

A drum, 2, of density $2000\,\text{kg}\,\text{m}^{-3}$, length $0.1\,\text{m}$ and diameter $0.3\,\text{m}$, is attached by way of a rigid connection to a block, 1, of mass $3\,\text{kg}$. The drum and block are on a plane inclined at 10° to the horizontal. The coefficient of friction between the two objects and the plane is 0.25. Calculate the acceleration of the system and the force in the connection.

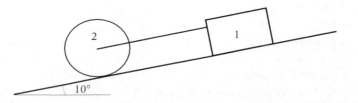

Solution

First consider the block. Draw a free-body diagram of the block, setting the force in the connection as T. From the free-body diagram it can be seen that it has been assumed that the connection is in tension. If this proves to be incorrect then the value obtained for T will be negative.

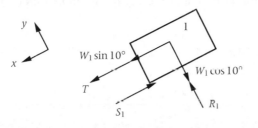

As usual, use translational equilibrium in the y-direction to determine R_1. Use the friction equation to establish S_1 and finally the translational motion equation in the y-direction to construct an equation in T and a_x.

$$\sum F_y = m_1 a_y = 0$$
$$R_1 - W_1 \cos 10° = 0$$
$$R_1 = 3 \times 9.81 \cos 10°$$
$$= 28.983 \text{ N}$$

$$S_1 = \mu R_1$$
$$= 0.25 \times 28.983$$
$$= 7.246 \text{ N}$$

$$\sum F_x = m_1 a_x$$
$$T + W_1 \sin 10° - S_1 = m_1 a_x$$
$$T = 7.246 - (3 \times 9.81 \sin 10°) + 3a_x$$
$$T = 2.136 - 3a_x \qquad \text{(eqn 1)}$$

Now consider the drum. Draw the free-body diagram of the drum.

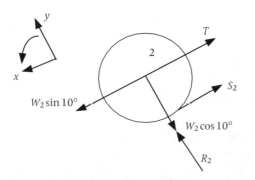

Assuming that the drum will be rolling down the slope rather than sliding, then:

$$S_2 \leq \mu R_2$$

First determine the mass and second moment of mass of the drum. Produce an equation in T, S_2 and a_x using the translational motion equation in the x-direction, and another in S_2 and α using the rotational motion equation.

$$
\begin{aligned}
m_2 &= \rho V \\
&= 2000\left(\pi \times 0.15^2 \times 0.1\right) \\
&= 14.137 \, \text{kg}
\end{aligned}
$$

$$
\begin{aligned}
I &= \frac{mr^2}{2} \\
I &= \frac{14.137 \times 0.15^2}{2} \\
&= 0.159 \, \text{kg m}^2
\end{aligned}
$$

$$\sum F_x = m_2 a_x$$
$$W_2 \sin 10° - T - S_2 = m_2 a_x$$
$$T = 14.137 \times 9.81 \sin 10° - S_2 - 14.137 a_x$$
$$T = 24.082 - S_2 - 14.137 a_x \qquad \text{(eqn 2)}$$

Note that only S_2 creates a moment about the mass centre since the other four forces have lines of action through the centre of the drum.

$$\sum M = I\alpha$$
$$0.15 S_2 = 0.159 \alpha$$
$$S_2 = 1.06 \alpha \qquad \text{(eqn 3)}$$

Since we are assuming that the drum is rolling without slipping we can rewrite eqn (3), replacing α by the equivalent value of a_x:

$$\alpha = \frac{a_x}{r}$$

$$\alpha = \frac{a_x}{0.15} \qquad \text{(eqn 4)}$$

$$S_2 = \frac{1.06 a_x}{0.15}$$
$$S_2 = 7.067 a_x$$

Substituting for S_2 in eqn (2):

$$T = 24.082 - 7.067 a_x - 14.137 a_x$$
$$T = 24.082 - 21.204 a_x \qquad \text{(eqn 5)}$$

Solving eqns (1) and (5) simultaneously:

$$24.082 - 21.204a_x = 2.136 - 3a_x$$

$$a_x = \frac{24.082 - 2.136}{18.204}$$

$$a_x = 1.206\ \mathrm{m\,s^{-2}}$$

The positive sign indicates the direction of the acceleration. Hence it is $1.206\ \mathrm{m\,s^{-2}}$ down the slope. The rotational acceleration of the drum can be obtained from eqn (4):

$$\alpha = \frac{a_x}{0.15}$$

$$\alpha = \frac{1.206}{0.15}$$

$$\alpha = 8.04\ \mathrm{rad\,s^{-2}}$$

From the axis system established on the free-body diagram, the positive sign indicates that the direction of the rotational acceleration is anticlockwise. The force, T, in the connection can be found from eqn (5):

$$T = 24.082 - (21.204 \times 1.206)$$

$$T = -1.49\ \mathrm{N}$$

The negative sign indicates that the original assumption that the connection was a tie was incorrect and it is in fact a strut carrying a force of $1.49\,\mathrm{N}$. Notice also that the minimum value of μ which will ensure that the drum rotates, as assumed, rather than slides can readily be calculated as follows:

$$R_2 = W_2 \cos 10°$$

$$R_2 = (14.137 \times 9.81)\cos 10°$$

$$R_2 = 39.967\ \mathrm{N}$$

$$\mu \geq \frac{S_2}{R_2}$$

$$\mu \geq \frac{1.06\alpha}{R_2}$$

$$\mu \geq \frac{1.06 \times 8.04}{39.967}$$

$$\mu \geq 0.21$$

The problem stated that μ was 0.25 which is greater than this minimum value and hence we were justified in assuming at the outset that the drum would roll and not slide.

EXAMPLE 4.6

A cylinder of mass m (kg) and radius r (m) is pulled over a horizontal surface by a force T (N) applied through a wire wrapped around its circumference. Assuming that the cylinder will roll, develop an equation which could be used to determine the force T for any given acceleration a_x and show that the resulting friction force developed between the cylinder and the surface is $T/3$.

Solution

Draw the free-body diagram and develop the translational and rotational equations of motion.

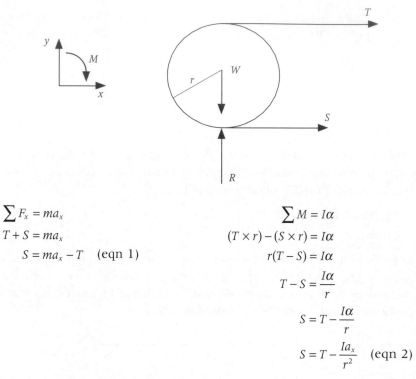

$$\sum F_x = ma_x$$

$$T + S = ma_x$$

$$S = ma_x - T \quad \text{(eqn 1)}$$

$$\sum M = I\alpha$$

$$(T \times r) - (S \times r) = I\alpha$$

$$r(T - S) = I\alpha$$

$$T - S = \frac{I\alpha}{r}$$

$$S = T - \frac{I\alpha}{r}$$

$$S = T - \frac{Ia_x}{r^2} \quad \text{(eqn 2)}$$

Solve the resulting two equations simultaneously to eliminate S, and hence develop the required relationship.

$$ma_x - T = T - \frac{Ia_x}{r^2}$$

$$2T = ma_x + \frac{Ia_x}{r^2}$$

$$T = \frac{a_x}{2}\left(m + \frac{I}{r^2}\right)$$

$$T = \frac{a_x}{2}\left(m + \frac{\dfrac{mr^2}{2}}{r^2}\right)$$

$$T = \frac{a_x}{2}\left(m + \frac{m}{2}\right)$$

$$T = \frac{3ma_x}{4} \qquad \text{(eqn 3)}$$

Substitute for T in eqn (1) and develop the corresponding equation for the friction force, S.

$$S = ma_x - \frac{3ma_x}{4}$$

$$S = \frac{ma_x}{4} \qquad \text{(eqn 4)}$$

Comparing eqn (3) and eqn (4) it can be seen that S equals $T/3$ for any mass, m, or acceleration, a_x.

4.2 Further discussion on Examples 4.1 to 4.6

These six examples further illustrate the importance of constructing free-body diagrams. The solution technique in each case involves first identifying and isolating the moving elements of the system, the movements of the individual elements being either translational, rotational or a combination of both. In Example 4.1 we have a single element, the drum, which is purely rotating. In Example 4.2 there are three elements, two of which are in pure translation and one in pure rotation. In Example 4.3 there is a single element, the wheel, which rotates and translates, and in Example 4.4 there are three elements, one in pure translation, one in pure rotation and one rotating and translating. In Example 4.5 there are two elements, one in pure translation while the other translates and rotates. Finally, Example 4.6 develops the general solution for a cylinder pulled along a horizontal surface by a wire wrapped around its circumference when the force in the wire is parallel to the surface.

In each case the solution method follows a similar pattern: construct equations for each element using Newton's second law, translational or rotational form as appropriate, and solve the resulting simultaneous equations to determine the unknowns.

4.3 Problems

4.1 The drum shown has a diameter of 0.25 m, a length of 0.5 m and a density of 8000 kg m⁻³. The drum rotates about a fixed bearing at A. Determine the rotational acceleration of the drum and the force at the bearings.

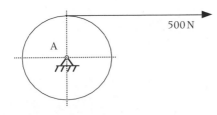

4.2 The concrete roller shown has a diameter of 0.75 m, a length of 1 m and a density of 2400 kg m^{-3}. The roller is being pushed along a horizontal surface by a force of 400 N directed through the axle centre and at an angle of 30° to the surface. Calculate the translational and rotational accelerations of the roller.

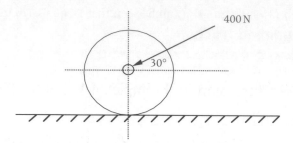

4.3 A 4 kg mass is connected by a wire passing over a small pulley to a drum. The drum has a diameter of 0.4 m, a length of 0.6 m and a density of 2000 kg m^{-3}. The wire is wound round the circumference of the drum. Determine the acceleration of the mass and the tension in the connecting wire.

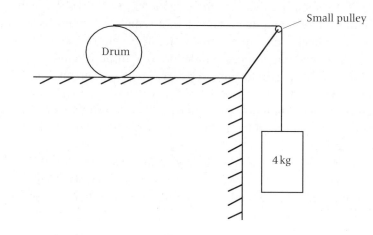

4.4 Rotational equilibrium

In Chapter 3 we saw that when a system is in translational equilibrium:

$$\sum F = 0 \tag{4.3}$$

Similarly if a system is not rotating there is no rotational acceleration. The system is in rotational equilibrium and hence:

$$\sum M = 0 \tag{4.4}$$

The sum of the moments acting on the system, about any point on the system is zero.

These two equilibrium conditions form the basis for the analysis of a wide range of statics problems.

4.5 Determining reactions

In many statics problems, initially only the applied loads will be known. These applied loads have to be resisted by reaction forces at the supports. The magnitude and direction of these reaction forces can be determined by applying the translational and rotational equilibrium equations.

EXAMPLE 4.7

Determine the support reactions at A and B on the 10.5 m span, simply supported beam, AB, subjected to the three applied point loads shown.

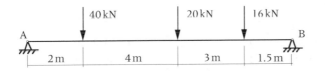

Solution

Draw the free-body diagram. Although the fixings at A and B would normally be represented by horizontal and vertical components, in this case, since the applied forces are all vertical, there can be no horizontal components and hence these will be omitted from the start.

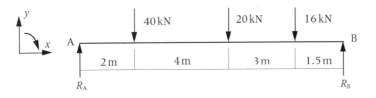

Select B as the appropriate point on the beam to which to apply the rotational equilibrium equation. Adopt the convention that clockwise moments are positive. The resulting equation only contains one unknown, R_A, since R_B produces no moment at B. The equation can then be solved to give the reaction force R_A.

$$\sum M_{\mathrm{B}} = 0$$

$$(10.5R_{\mathrm{A}}) - (40 \times 8.5) - (20 \times 4.5) - (16 \times 1.5) = 0$$

$$R_{\mathrm{A}} = \frac{340 + 90 + 24}{10.5}$$

$$R_{\mathrm{A}} = 43.238 \text{ kN}$$

The reaction, R_{A}, has a magnitude of 42.238 kN. The answer is positive, indicating that the direction chosen on the free-body diagram for the force is correct, i.e. ↑.

The reaction force R_{B} could be found in a similar manner by summing the moments about point A. Alternatively the translational equilibrium equation can be used:

$$\sum F_{y} = 0$$

$$42.238 - 40 - 20 - 16 + R_{\mathrm{B}} = 0$$

$$R_{\mathrm{B}} = 33.762 \text{ N}$$

Hence the reaction force, R_{B}, is 33.762 N ↑.

EXAMPLE 4.8

Determine the support reactions, R_{A} and R_{B}, on the beam shown below. In this, and all following examples, where the applied loading is purely vertical, we will adopt the normal short-cut of describing the problem directly as a free-body diagram.

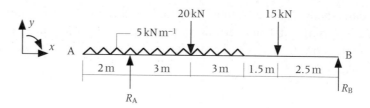

Solution

By taking moments about the right-hand end of the beam the resulting equilibrium equation can be solved for R_{A}.

$$\sum M_{\mathrm{RB}} = 0$$

$$10R_{\mathrm{A}} - (20)(7) - (15)(2.5) - (5)(8)(8) = 0$$

$$R_{\mathrm{A}} = \frac{140 + 37.7 + 320}{10}$$

$$R_{\mathrm{A}} = 49.75 \text{ kN}$$

The reaction R_{A} is therefore 49.75 kN ↑.

$$\sum F_y = 0$$
$$49.75 + R_B - 20 - 15 - (5)(8) = 0$$
$$R_B = 25.25 \text{ kN}$$

The reaction R_B is therefore 25.25 kN $\uparrow$.

EXAMPLE 4.9

Determine the support reactions at A and C on the pin-jointed frame shown below. This frame was first introduced in Chapter 3, Section 3.7, 'Further discussion on Examples 3.6 to 3.9'.

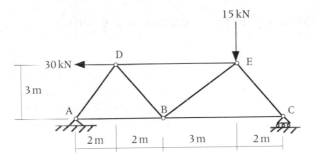

Solution

First draw the free-body diagram of the frame, noting that the fixing at A provides vertical and horizontal restraint while the fixing at C provides vertical restraint only.

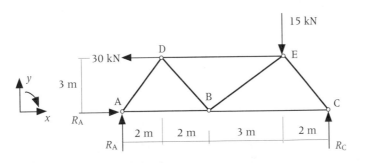

By taking moments about A, we can set up an equation which can be solved to give R_C. Take care to multiply forces by their perpendicular distance to A. The perpendicular distance from both R_C and the 15 kN load to A are horizontal measurements, but the perpendicular distance from the 30 kN load to A is a vertical measurement.

$$\sum M_A = 0$$

$$(15)(7) - (30)(3) - 9R_C = 0$$

$$R_C = \frac{105 - 90}{9}$$

$$R_C = 1.667 \text{ kN} \uparrow$$

Use vertical equilibrium to determine R_{AV} and horizontal equilibrium to determine R_{AH}:

$$\sum F_y = 0 \qquad\qquad \sum F_x = 0$$

$$R_{AV} + 1.667 - 15 = 0 \qquad R_{AH} - 30 = 0$$

$$R_{AV} = 13.333 \text{ kN} \uparrow \qquad R_{AH} = 30 \text{ kN} \rightarrow$$

4.6 | Pin-jointed frames: method of sections

$\sum M = 0$, $\sum F_x = 0$ and $\sum F_y = 0$ are true not only for the complete system, but also hold for any subsystem. If we were to isolate a section of a frame, and draw the free-body diagram of the section of frame, then this would itself be in equilibrium. By considering an appropriate sub-frame, it is possible to determine member forces by the technique known as the *method of sections*.

EXAMPLE 4.10

Using the method of sections, determine the forces in the members CD, CB and AB of the pin-jointed frame first analysed in Example 3.9.

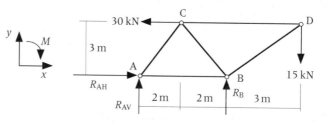

Solution

It is best to first determine the external reaction forces. If we take moments about A, then by rotational equilibrium, we can set up an equation which contains R_B as the only unknown.

$$\sum M_A = 0$$

$$(15 \times 7) - (30 \times 3) - 4R_B = 0$$

$$R_B = \frac{(15 \times 7) - (30 \times 3)}{4}$$

$$R_B = 3.75 \text{ kN}$$

Use translational equilibrium to determine the horizontal and vertical components of the reaction force at A:

$$\sum F_x = 0 \qquad\qquad \sum F_y = 0$$

$$R_{AH} - 30 = 0 \qquad\qquad R_{AV} + 3.75 - 15 = 0$$

$$R_{AH} = 30 \text{ kN} \qquad\qquad R_{AV} = 11.25 \text{ kN}$$

Consider the equilibrium of the sub-frame selected. Taking moments about C produces an equation which can be solved for F_{AB}.

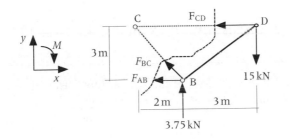

$$\sum M_C = 0$$

$$(15 \times 5) - (3.75 \times 2) + 3F_{AB} = 0$$

$$F_{AB} = \frac{(3.75 \times 2) - (15 \times 5)}{3}$$

$$F_{AB} = -22.5 \text{ kN}$$

On the free-body diagram of the sub-frame AB was assumed to be a tie. The negative result obtained indicates that it is in fact a strut. Now take moments about B to determine F_{CD}.

$$\sum M_B = 0$$

$$(15 \times 3) - 3F_{CD} = 0$$

$$F_{CD} = 15 \text{ kN}$$

The positive result indicates that the member is a tie as assumed. Finally to determine F_{BC} we need to split F_{BC} into its horizontal and vertical components and then use translational equilibrium in the y-direction.

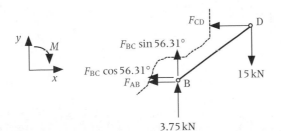

$$\sum F_y = 0$$

$$3.75 + F_{BC} \sin 56.31° - 15 = 0$$

$$F_{BC} = \frac{15 - 3.75}{\sin 56.31°}$$

$$F_{BC} = 13.52 \text{ kN}$$

So BC is a tie carrying a force of 13.52 kN. Compare the results with those obtained in Chapter 3 using the method of joints. The slight differences are due to rounding errors.

4.7 Further discussion on Examples 4.7 to 4.10

Examples 4.7 to 4.9 show how the equilibrium forms of Newton's second law for translation and rotation can be used to determine reaction forces in statics problems. These same two equations can be used in the analysis of complete systems or equally well applied to isolated sections of a system. Example 4.10 shows how this approach can be used when determining the forces in the members of a pin-jointed frame. This approach, known as the *method of sections* is more efficient than the method of joints technique described in Chapter 4 if we are to determine the force in a central member of a large frame, since this can be done directly rather than gradually progressing from pin to pin.

4.8 Problems (continued)

In **4.4** to **4.7**, determine the support reactions, R_A and R_B.

4.4

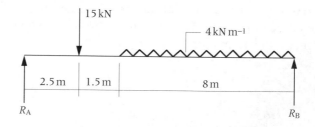

4.5

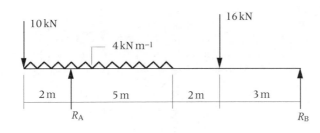

4.6

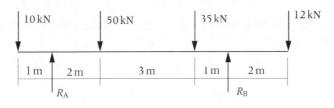

4.7

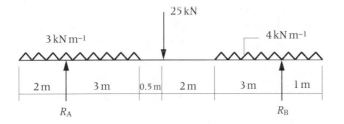

4.8   Using the method of sections, determine the forces in members CD, CB and AB of the pin-jointed frame shown. Compare the answers with those obtained in **3.11** of Chapter 3.

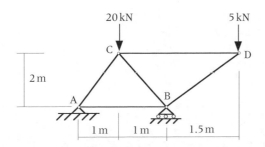

4.9 Using the method of sections, determine the forces in members AC, AD, BD and AB of the pin-jointed frame shown.

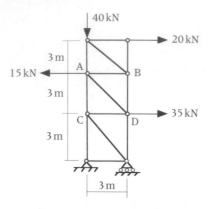

4.10 Using the method of sections, determine the forces in members DE, DB and BC of the pin-jointed frame shown.

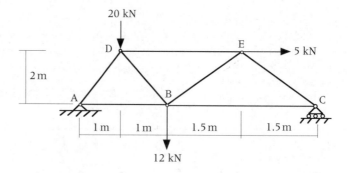

CHAPTER 5

Work–Energy Methods

Chapters 3 and 4 illustrated the use of Newton's laws of motion in the solution of a large range of problems in engineering mechanics. In certain situations a more appropriate approach is through the application of mechanical work and energy. These concepts, and the powerful work–energy method, are introduced in this chapter. Initially work and energy are considered separately with full worked examples given, showing how these fundamental quantities can be calculated. The relationship between work and energy is then discussed and the work–energy equation is outlined. Typical examples are then given of problems where a work–energy approach to their solution is appropriate, and these are worked through in detail.

The following concepts are introduced in this chapter:

- *Work*
- *Energy*
- *Kinetic energy*

- *Strain energy*
- *Potential energy*

5.1 | Work

Work is defined as the mechanical means of transferring energy from one body to another. Work is done by forces and moments acting on objects and causing movement, and is a scalar.

$$\Delta = Fs$$ (5.1)

where Δ = work done by a force (Nm)
  F = force (N)
  s = distance travelled in the direction of the force (m).

$$\Delta = M\theta$$ (5.2)

where M = applied moment or torque (Nm)
  θ = angle of rotation (rad).

In problems involving work and energy, 1 Nm is defined as 1 Joule (J).

EXAMPLE 5.1

A 100 kg crate is pushed a distance of 4 m across a horizontal floor by a force of 500 N. If the coefficient of friction between the crate and the floor is 0.4, determine the total work done on the crate.

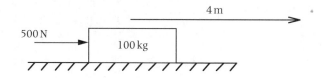

Solution

First draw a free-body diagram to determine the forces acting on the crate.

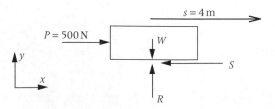

Note that the 500 N force is in the direction of motion of the crate and therefore does positive work on the crate whereas the friction force, S, opposes the direction of motion and hence does negative work. The normal reaction, R, and the weight force, W, are not applied in the direction of motion and do no work on the crate. We need to determine the friction force, S, in the usual way.

$$\sum F_y = ma_y = 0$$
$$R - W = 0$$
$$R = W$$
$$R = 100 \times 9.81$$
$$= 981\,\text{N}$$

$$F = \mu R$$
$$= 0.4 \times 981$$
$$= 392.4\,\text{N}$$

We can now determine the individual components of work done on the crate and sum these to give the total work done on the crate.

$$\Delta = Fs$$
$$\Delta_P = (500)(4) = 2000\,\text{J}$$
$$\Delta_F = (-392.4)(4) = -1570\,\text{J}$$
$$\Delta_W = 0$$
$$\Delta_R = 0$$
$$\therefore \Delta = 2000 - 1570 = 430\,\text{J}$$

Hence the total work done on the crate is 430 J.

EXAMPLE 5.2

Calculate the work done on the trap-door shown in order to rotate it about the hinge at A, through 90°. The 80 N force maintains an angle of 60° to the door throughout the rotation.

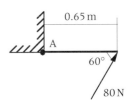

Solution

Draw the free-body diagram and calculate the work done by determining the moment of the forces about the hinge, A.

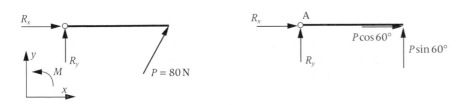

The only force causing a moment about A is the vertical component of P. Remember that the angle of rotation, θ, must be in radians.

$$\Delta = M\theta$$

$$\Delta = P \sin\theta \times 0.65$$

$$\Delta = (80 \sin 60°)\left(\frac{90 \times \pi}{180}\right)(0.65)$$

$$= 70.738 \text{ J}$$

Hence the work done on the trap-door is 70.738 J.

EXAMPLE 5.3

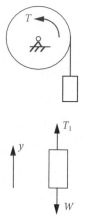

A drum has a diameter of 0.6 m, a length of 1 m and density of 2430 kg m⁻³. The drum is driven by an anticlockwise torque, T, causing an angular acceleration of 0.5 rad s⁻² while raising a 90 kg mass. Determine the work done on the mass and the work done on the drum during 3 revolutions.

Solution

First draw a free-body diagram of the mass.

Since we are given the rotational acceleration of the drum we can calculate the upwards translational acceleration of the mass:

$$a_{mass} = \alpha r_{drum}$$
$$= (0.5)(0.3)$$
$$= 0.15 \text{ ms}^{-2} \uparrow$$

We can also calculate the vertical distance travelled by the mass during the 3 revolutions:

$$S = 3\pi d$$
$$= 3\pi(0.6)$$
$$= 5.654 \text{ m}$$

Use Newton's second law to determine the tension in the wire, T_1:

$$\sum F_y = ma_y$$
$$T_1 - W = ma_y$$
$$T_1 = W + ma_y$$
$$= (90 \times 9.81) + (90 \times 0.15)$$
$$= 896.4 \text{ N}$$

Now determine the work done on the mass by the individual forces, T_1 and W, remembering that we have defined vertically upwards as the positive direction of motion on the free-body diagram.

$$\Delta = F \times S$$
$$\Delta_{T_1} = (896.4)(5.654) = 5068 \text{ J}$$
$$\Delta_W = -(90 \times 9.81)(5.654) = -4992 \text{ J}$$
$$\therefore \Delta = 5068 - 4992 = 76 \text{ J}$$

Hence the total work done on the mass during the 3 revolutions of the drum is 76 J.

To determine the work done on the drum we first need a free-body diagram of the drum.

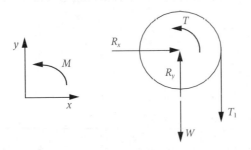

To determine the value of the input torque, T, we will need to apply the rotational form of Newton's second law. We therefore first require to calculate the second moment of mass of the winding drum.

$$m = \rho V$$
$$= (2400)(\pi \times 0.3^2 \times 1)$$
$$= 678.584 \text{ kg}$$

$$I = \frac{mr^2}{2}$$
$$= \frac{(678.584)(0.3)^2}{2}$$
$$= 30.536 \text{ kg m}^2$$

Now apply Newton's second law:

$$\sum M = I\alpha$$
$$T - (T_1 r) = I\alpha$$
$$T = (T_1 r) + (I\alpha)$$
$$= (896.4 \times 0.3) + (30.536 \times 0.5)$$
$$= 284.188 \text{ N m}$$

Positive work is done on the drum by torque T, and negative work is done by the T_1 force. Note that the forces W, R_x and R_y do not contribute to the rotation of the drum and therefore do no work.

$$\Delta = (F \times S) + (M\theta)$$
$$= -(896.4 \times 5.654) + (284.188)\left(\frac{3 \times 360 \times \pi}{180} \right)$$
$$= -5068 + 5357$$
$$= 289 \text{ J}$$

The total work done on the drum during the 3 revolutions is 289 J.

5.2 Energy

Mechanical energy is the ability that an object or system has to do work. Work is therefore the mechanical means of transmission of energy from one object to another.

$$E_2 = E_1 + \Delta \tag{5.3}$$

where E_2 = the final energy in the system (J)
E_1 = the initial energy in the system (J)
Δ = the work done on the system (J).

Mechanical energy has two components: kinetic energy, K, which is the energy of motion, and strain energy, S, the energy of deformation.

$$E = K + S \tag{5.4}$$

5.3 Kinetic energy

An object has kinetic energy due to its translational and rotational motions.

$$K = \frac{1}{2}mv^2 \quad \text{translation}$$
$$K = \frac{1}{2}I\omega^2 \quad \text{rotation} \tag{5.5}$$

where m = the object's mass (kg)
v = the object's translational velocity (m s^{-1})
I = the object's second moment of mass (kg m^2)
ω = the object's angular velocity about the mass centre (rad s^{-1}).

EXAMPLE 5.4

Determine the kinetic energy of an 80 kg crate moving with a velocity of 1.5 m s^{-1} over a horizontal surface.

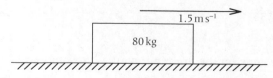

Solution

The crate has translational kinetic energy:

$$K = \frac{1}{2}mv^2$$
$$= \frac{(80)(1.5)^2}{2}$$
$$= 90\,\text{J}$$

The kinetic energy of the crate at the instant shown is 90 J.

EXAMPLE 5.5

A wheel with a diameter of 0.6 m and 0.1 m thickness rotates about its centre with an angular velocity of $100\,\text{rad}\,\text{s}^{-1}$. The wheel has a mass of 75 kg. Determine the kinetic energy.

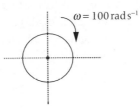

Solution

The wheel has rotational kinetic energy. To calculate this we first need to determine the second moment of mass of the wheel.

$$I = \frac{mr^2}{2}$$
$$= \frac{(75)(0.3)^2}{2}$$
$$= 3.375\,\text{kg}\,\text{m}^2$$

$$K = \frac{1}{2}I\omega^2$$
$$= \frac{(3.375)(100)^2}{2}$$
$$= 16\,875\,\text{J}$$
$$= 16.875\,\text{kJ}$$

The kinetic energy of the wheel at the instant shown is 16.875 kJ.

5.4 | Strain energy

Strain energy is the energy of deformation. Forces applied to objects cause shape changes. For most objects and situations these changes are small and can be neglected. For some materials, however, such as rubber or springs, the shape changes can be significant.

If a spring with a linear spring stiffness k ($\mathrm{Nm^{-1}}$) is extended or compressed by an amount e (m) through the application of a force P (N), then the work done on the spring is equal to the area under the load–extension graph.

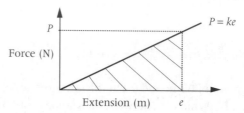

Notice that in problems involving strain energy, the force P is proportional to the deformation occurring. When a force or forces acting on a system varies, so also will the acceleration and we are not then dealing with a constant acceleration model and consequently the kinematic formulae from Chapter 2 will not apply.

$$\Delta = \int_0^e ke \cdot \mathrm{d}e$$

$$= \frac{ke^2}{2}$$

Hence the strain energy that is stored in the system is given by the following formula:

$$S = \frac{1}{2}ke^2 \qquad\qquad (5.6)$$

where k = the system stiffness ($\mathrm{Nm^{-1}}$)

e = the deformation from the free length (m).

The deformation can be a contraction or extension but always takes a positive value.

5.5 | Potential energy

You may already have come across the term 'potential energy'. Potential energy is defined as the energy in a system due to its mass, m (kg), and elevation, h(m), above a convenient datum, and is calculated as mgh (J). In the work–energy method presented in this chapter, potential energy is not calculated. Instead the effect of changes in the potential energy, as the relative vertical position of the system alters, is included as a work term with the weight force treated in the same way as any other force acting on the system. The following illustration shows that the two approaches to treating weight are equivalent.

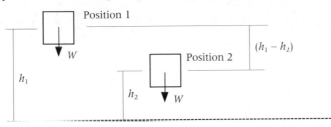

As the system moves from position 1 to position 2, the potential energy changes from mgh_1 (J) to mgh_2 (J), a difference of $W(h_1 - h_2)$. If we were, instead, to calculate the work done by the weight force, we again obtain $W(h_1 - h_2)$. The only difference being that this time we view it as a work term rather than an energy term.

5.6 | Work–energy method

As was stated earlier, work is the mechanical means of transmission of energy from one object to another. The final energy in a system is simply equal to the initial energy (kinetic and strain) in the system plus the work done on the system (including the work done by the weight force). The fundamental work–energy equation is therefore:

$$E_2 = E_1 + \Delta \tag{5.7}$$

where E_2 = the final energy in the system (J)
$\quad$ E_1 = the initial energy in the system (J)
$\quad$ Δ = the work done on the system (J).

This basic equation provides a means of solving a range of dynamics questions. Firstly we will look at an example where the energy component arises purely from motion – there is no deformation taking place. We therefore have kinetic energy present but no strain energy.

EXAMPLE 5.6

A 20 kg crate slides down a 30° slope under the influence of its own weight. The coefficient of friction between the crate and the sloping surface is 0.2. At an instant the speed of the crate is found to be $1\,\mathrm{m\,s^{-1}}$. Determine the crate's speed 6 m further down the slope.

Solution

First draw a sketch of the problem and a free-body diagram.

Determine the reaction and friction forces as usual:

$$\sum F_y = ma_y$$

$$R - W\cos 30° = 0$$

$$R = (20 \times 9.81)\cos 30°$$

$$R = 169.914 \text{ N}$$

$$S = \mu R$$

$$= 0.2 \times 169.914$$

$$= 33.983 \text{ N}$$

Now calculate the work done during the 6 m movement down the slope. Note that only the component of weight force parallel to the slope and the friction force do any work. The friction force does negative work on the crate.

$$\Delta = Fs$$

$$= (W\sin 30° \times 6) - (S \times 6)$$

$$= (20 \times 9.81\sin 30° \times 6) - (33.983 \times 6)$$

$$= 384.702 \text{ J}$$

Calculate the initial and final kinetic energy:

$$K_1 = \frac{mv_1^2}{2}$$

$$= \frac{(20)(1)^2}{2}$$

$$= 10 \text{ J}$$

$$K_2 = \frac{mV_2^2}{2}$$

$$= \frac{20V_2^2}{2}$$

$$= 10V_2 \text{ J}$$

Now apply the work–energy equation. The energy components are purely kinetic. The unknown, V_2, can then be obtained by rearranging the equation.

$$E_2 = E_1 + \Delta$$

$$K_2 = K_1 + \Delta$$

$$10V_2^2 = 10 + 384.702$$

$$V_2 = \sqrt{\frac{394.702}{10}}$$

$$= 6.283 \text{ m s}^{-1}$$

The velocity of the crate after travelling 6 m down the slope is 6.283 ms^{-1}.

The same result could of course have equally well been obtained by using Newton's method detailed in Chapter 3 along with the kinematic equations for constant acceleration developed in Chapter 2.

$$\sum F_x = ma_x$$

$$W \sin 30° - S = ma_x$$

$$a_x = \frac{(20 \times 9.81 \times \sin 30°) - 33.983}{20}$$

$$= 3.206 \text{ m s}^{-2}$$

$$v^2 = u^2 + 2ax$$

$$v = \sqrt{(1)^2 + (2)(3.206)(6)}$$

$$v = 6.283 \text{ m s}^{-1}$$

As before, the velocity of the crate after travelling 6 m down the slope is found to be 6.283 ms^{-1}.

EXAMPLE 5.7

In this example, translational kinetic energy and strain energy are relevant. A truck with a mass of 25×10^3 kg is travelling at 2 ms^{-1}. The truck is fitted with a buffer spring with a stiffness 900 kN m^{-1}. (a) Determine the compression in the spring if the truck is brought to an immediate stop on collision with a solid block. (b) Also if the maximum permissible compression of the spring is 0.5 m then calculate the maximum safe speed of arrival.

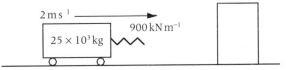

Solution

(a) Determine the energy of the truck prior to the collision. The truck has kinetic energy due to its speed but the strain energy is zero as the spring is in its equilibrium position.

$$E_1 = K_1 + S_1$$

$$= \frac{mv_1^2}{2} + 0$$

$$= \frac{25\,000 \times 2^2}{2} + 0$$

$$= 50\,000 \text{ J}$$

Now determine the energy of the truck after the collision. The kinetic energy is zero since the truck is at rest but the spring has gained strain energy due to its compression.

$$E_2 = K_2 + S_2$$

$$= 0 + \frac{ke^2}{2}$$

$$= 0 + \frac{900\,000e^2}{2}$$

$$= 450\,000e^2 \text{ J}$$

Determine the work done. If we assume the stop to be instantaneous then although there will be considerable forces acting, the distance travelled during the action of these forces is zero so they do no work.

$$\Delta = Fs$$

$$= (F)(0)$$

$$= 0$$

Now use the work–energy equation.

$$E_2 = E_1 + \Delta$$

$$450\,000e^2 = 50\,000 + 0$$

$$e = \sqrt{\frac{50\,000}{450\,000}}$$

$$= 0.333 \text{ m}$$

The spring is compressed by 0.333 m to bring the truck to an instantaneous stop.

(b) In this part of the question the unknown is v_1.

$$E_1 = K_1 + S_1$$

$$= \frac{mv_1^2}{2} + 0$$

$$= \frac{25\,000\,v_1^2}{2}$$

$$= 12\,500v_1^2 \text{ J}$$

$$E_2 = K_2 + S_2$$

$$= 0 + \frac{ke^2}{2}$$

$$= 0 + \frac{900\,000 \times 0.5^2}{2}$$

$$= 112\,500 \text{ J}$$

$$E_2 = E_1 + \Delta$$

$$112\,500 = 12\,500v_1^2 + 0$$

$$v_1 = \sqrt{\frac{112\,500}{12\,500}}$$

$$= 3 \text{ m s}^{-1}$$

The maximum approach speed to limit the compression of the spring to 0.5 m is $3\,\text{m s}^{-1}$.

EXAMPLE 5.8

This example involves both rotational and translational kinetic energy but there is no deformation and therefore no strain energy. A 0.4 m diameter cylinder with a mass of 8 kg rolls 3 m down a 20° slope from a standing start. Determine the minimum value for the coefficient of friction between the two surfaces to ensure that rolling rather than sliding takes place, and calculate the final values of the translational and angular velocities of the cylinder.

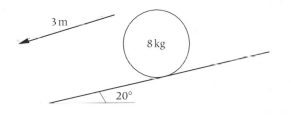

Solution

(a) Begin by drawing the free-body diagram, then use the translational form of Newton's second law to determine the reaction force, R, and an equation relating the friction force, S, and the translational acceleration down the slope, a_x. Remember that if rolling takes place then $S \le \mu R$. We can only set $S = \mu R$ if we have sliding.

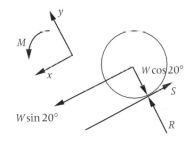

$$\sum F_y = ma_y = 0$$
$$R - W\cos 20° = 0$$
$$R = (8 \times 9.81)\cos 20°$$
$$R = 73.747 \text{ N}$$

$$\sum F_x = ma_x$$
$$W\sin 20° - S = ma_x$$
$$S = (8 \times 9.81 \times \sin 20°) - 8a_x$$
$$S = 26.842 - 8a_x \quad \text{(eqn 1)}$$

Set up a second equation, this time using the rotational form of Newton's second law, but first you need to calculate the second moment of mass of the cylinder.

$$I = \frac{mr^2}{2}$$

$$= \frac{(8)(0.2)^2}{2}$$

$$= 0.16 \, \text{kg m}^2$$

$$\sum M = I\alpha$$

$$0.2S = 0.16\alpha$$

$$S = 0.8\alpha \quad \text{(eqn 2)}$$

We can rewrite eqn (2) in terms of S and a_x:

$$a_x = \alpha r$$

$$a_x = 0.2\alpha$$

$$\alpha = \frac{a_x}{0.2}$$

$$\therefore S = 0.8 \times \frac{a_x}{0.2}$$

$$S = 4a_x \quad \text{(eqn 3)}$$

Solve eqns (1) and (3) simultaneously to determine a_x and hence by substitution α.

$$4a_x = 26.842 - 8a_x$$

$$12a_x = 26.842$$

$$a_x = 2.237 \, \text{m s}^{-2}$$

$$\alpha = \frac{2.237}{0.2}$$

$$\alpha = 11.185 \, \text{rad s}^{-2}$$

From eqn (3) it is now possible to calculate the friction force, S:

$$S = 4a_x$$

$$= (4)(2.237)$$

$$= 8.948 \, \text{N}$$

For rolling to take place $S \le \mu R$ and hence $\mu \ge \dfrac{8.948}{73.747}$, i.e. we need $\mu \ge 0.12$.

(b) The translational and rotational velocities of the cylinder at the foot of the 3 m slope can be obtained using the work–energy equation (but see also the direct method at the end of the example).

$$E_2 = E_1 + \Delta$$

The cylinder has both translational and rotational kinetic energy but no strain energy.

$$K_2 = K_1 + \Delta$$

Calculate the initial kinetic energy of the cylinder. It is of course zero since the cylinder is initially at rest.

$$K_1 = \frac{mv_1^2}{2} + \frac{I\omega_1^2}{2}$$
$$= 0 + 0$$

Now calculate the final kinetic energy of the cylinder in terms of the final translational velocity, v_2, and the final angular velocity, ω_2.

$$K_2 = \frac{mv_2^2}{2} + \frac{I\omega_2^2}{2}$$
$$= \frac{8v_2^2}{2} + \frac{0.16\omega_2^2}{2}$$
$$= 4v_2^2 + 0.08\omega_2^2 \quad \text{(eqn 4)}$$

Equation (4) can be written purely in terms of ω_2 by substituting $v = \omega r$.

$$K_2 = 4(\omega_2 r)^2 + 0.08\omega_2^2$$
$$K_2 = 0.24\omega_2^2 \quad \text{(eqn 5)}$$

Now compute the work done on the cylinder during the motion. It is important to note that the friction force, S, does no work during rolling.

$$\Delta = Fs$$
$$= (W \sin 20° \times 3) - (S \times 0)$$
$$= 8 \times 9.81 \times 3$$
$$= 80.525 \text{ J}$$

Finally calculate the final angular velocity from the work–energy equation:

$$K_2 = K_1 + \Delta$$
$$0.24\omega_2^2 = 0 + 80.825$$
$$\omega_2 = \sqrt{\frac{80.525}{0.24}}$$
$$= 18.317 \text{ rad s}^{-1}$$

Hence the final translational velocity:

$$v_2 = \omega_2 r$$
$$= 18.317 \times 0.2$$
$$= 3.663 \text{ m s}^{-1}$$

At the foot of the 3 m slope the cylinder will have a translational velocity of 3.663 m s^{-1} down the slope and an angular velocity of $18.317 \text{ rad s}^{-1}$ anticlockwise. Since we are dealing here with a constant acceleration model, we could have calculated the translational and angular velocities directly using kinematics. From part (a): $a_x = 2.237 \text{ m s}^{-2}$.

$$v^2 = u^2 + 2ax$$
$$v = \sqrt{(0)^2 + 2(2.237)(3)}$$
$$v = 3.663 \text{ m s}^{-1}$$

EXAMPLE 5.9

In this example we have translational kinetic energy, rotational kinetic energy and strain energy. At the instant shown for the system below, the spring, which has a stiffness of $800\,\mathrm{N\,m^{-1}}$, is compressed by $150\,\mathrm{mm}$ and block A has a velocity of $1.5\,\mathrm{m\,s^{-1}}\downarrow$. Block A has a mass of $60\,\mathrm{kg}$. Block C has a mass of $140\,\mathrm{kg}$. Wheel B has a mass of $87\,\mathrm{kg}$. Determine the velocity of block A after it has dropped $1.25\,\mathrm{m}$. The coefficient of friction between block C and the supporting surface is 0.3.

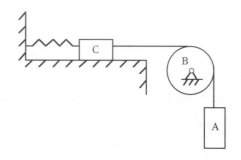

Solution

Begin by drawing a free-body diagram of the system.

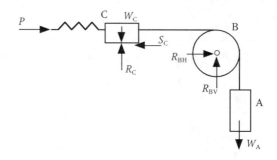

The force P is proportional to the deformation of the spring and hence in this example we do not have constant acceleration.

Calculate the initial kinetic energy and the initial strain energy:

$$K_1 = \frac{m_\mathrm{C}v_1^2}{2} + \frac{I_\mathrm{B}\omega_1^2}{2} + \frac{m_\mathrm{A}v_1^2}{2}$$

$$K_1 = \frac{m_\mathrm{C}v_1^2}{2} + \frac{\dfrac{m_\mathrm{B}r^2}{2}\left(\dfrac{v_1}{r}\right)^2}{2} + \frac{m_\mathrm{A}v_1^2}{2}$$

$$K_1 = \frac{m_\mathrm{C}v_1^2}{2} + \frac{m_\mathrm{B}v^2}{2} + \frac{m_\mathrm{A}v_1^2}{2}$$

$$K_1 = \left(\frac{140\times1.5^2}{2}\right) + \left(\frac{87\times1.5^2}{4}\right) + \left(\frac{60\times1.5^2}{2}\right)$$

$$K_1 = 273.938\,\mathrm{J}$$

$$S_1 = \frac{ke_1^2}{2}$$

$$S_1 = \frac{800 \times 0.15^2}{2}$$

$$S_1 = 9 \text{ J}$$

The initial energy of the system is the sum of the kinetic and strain energies:

$$E_1 = K_1 + S_1$$

$$E_1 = 273.938 + 9$$

$$E_1 = 282.938 \text{ J}$$

Now determine the expression for the final kinetic energy of the system:

$$K_2 = \frac{m_C v_2^2}{2} + \frac{I_B \omega_2^2}{2} + \frac{m_A v_2^2}{2}$$

$$K_2 = \frac{m_C v_2^2}{2} + \frac{\frac{m_B r^2}{2} \left(\frac{v_2}{r}\right)^2}{2} + \frac{m_A v_2^2}{2}$$

$$K_2 = \frac{m_C v_2^2}{2} + \frac{m_B v_2^2}{4} + \frac{m_A v_2^2}{2}$$

$$K_2 = \left(\frac{140 \times v_2^2}{2}\right) + \left(\frac{87 \times v_2^2}{4}\right) + \left(\frac{60 \times v_2^2}{2}\right)$$

$$K_2 = 121.750 v_2^2 \text{ J}$$

Determine the final strain energy of the system:

$$S_2 = \frac{ke_2^2}{2}$$

$$S_2 = \frac{800 \times (1.25 - 0.15)^2}{2}$$

$$S_2 = \frac{800 \times 1.1^2}{2}$$

$$S_2 = 484 \text{ J}$$

The final energy of the system is the sum of the final kinetic and strain energies:

$$E_2 = K_2 + S_2$$

$$E_2 = \left(121.750 v_2^2 + 484\right) \text{ J}$$

Now determine the work done on the system. There is a contribution from the weight of block A and a negative contribution from the friction force acting along the base of block C. None of the other forces present move and therefore do no work.

$$\Delta = Fs$$

$$\Delta = (W_A)s - (S_C)s$$

$$\Delta = (m_A g)s - (\mu m_C g)s$$

$$\Delta = (60 \times 9.81)1.25 - (0.3 \times 140 \times 9.81)1.25$$

$$\Delta = 220.725 \text{ J}$$

We can now equate the expressions for the initial energy, final energy and work done. The unknown velocity v_2 can then be calculated.

$$E_2 = E_1 + \Delta$$
$$121.7505v_2^2 + 484 = 282.938 + 220.725$$
$$v_2 = \sqrt{\frac{282.938 + 220.725 - 484}{121.750}}$$
$$v_2 = 0.402 \text{ m s}^{-1}$$

The velocity of block A after dropping $1.25\,\text{m}$ is $0.402\,\text{m s}^{-1}\ \downarrow$.

EXAMPLE 5.10

The spring in the system shown has a stiffness of $255\,\text{N m}^{-1}$ and a free length of $0.3\,\text{m}$. The spring is attached to a block which has a mass of $4\,\text{kg}$. The block slides over a level surface and the coefficient of friction between the block and the surface is 0.35. If the spring is compressed to $0.2\,\text{m}$ and then the block released, determine the velocity of the block during the subsequent motion as the spring returns to its free length.

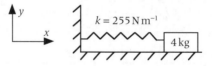

Solution

Draw a free-body diagram of the system and calculate the friction force, S.

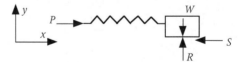

$$\sum F_y = 0$$
$$R - W = 0$$
$$R = W$$
$$R = 4 \times 9.81$$
$$R = 39.24 \text{ N}$$
$$S = \mu R$$
$$S = 0.35 \times 39.24$$
$$S = 13.734 \text{ N}$$

The compression of the spring, e, during the motion of the block will equal the free length minus the compressed length minus the block movement, x.

$$e = 0.3 - 0.2 - x$$
$$e = (0.1 - x)$$

Calculate the initial kinetic energy and strain energy of the system:

$$K_1 = \frac{mv_1^2}{2}$$
$$K_1 = 0$$
$$S_1 = \frac{ke_1^2}{2}$$
$$S_1 = \frac{(255)(0.1-0)^2}{2}$$
$$S_1 = 1.276 \text{ J}$$
$$E_1 = K_1 + S_1$$
$$E_1 = 1.276 \text{ J}$$

Calculate the kinetic and strain energies of the system after the block has moved a distance x (m):

$$K_x = \frac{mv_x^2}{2}$$
$$K_x = \frac{4v_x^2}{2}$$
$$K_x = 2v_x^2$$
$$S_x = \frac{ke_x^2}{2}$$
$$S_x = \frac{(255)(0.1-x)^2}{2}$$
$$S_x = 127.5(0.1-x)^2 \text{ J}$$
$$E_x = K_x + S_x$$
$$E_x = \left(2v_x^2 + 127.5(0.1-x)^2\right) \text{ J}$$

Determine an expression for the work done on the system during this movement, x, of the block.

$$\Delta = -Sx$$
$$\Delta = (-13.734x) \text{ J}$$

Equate the system energies and the equivalent work done and rearrange the resulting equation to make the velocity, v_x, the subject.

$$E_2 = E_1 + \Delta$$
$$2v_x^2 + 127.5(0.1-x)^2 = 1.275 - 13.734x$$
$$v_x = \sqrt{\frac{1.275 - 13.734x - 127.5(0.1-x)^2}{2}} \text{ m s}^{-1}$$

We can now use this expression to determine the velocity of the block, v_x, while incrementing x. The outcome is shown in the table overleaf:

Spring length (m)	Spring compression e (m)	Block position x (m)	Block velocity v_x (m s^{-1})
0.20	0.10	0	0
0.21	0.09	0.01	0.23
0.22	0.08	0.02	0.30
0.23	0.07	0.03	0.35
0.24	0.06	0.04	0.37
0.25	0.05	0.05	0.37
0.26	0.04	0.06	0.35
0.27	0.03	0.07	0.32
0.28	0.02	0.08	0.25
0.29	0.01	0.09	0.11
0.30	0	0.10	–

The equation has no real solution at x equal to 0.10 m, which would represent the spring returning to its free length. Hence we conclude that the block stops moving before this. Some compression will remain in the spring, generating a force which is insufficient to overcome the friction force.

The inclusion of the spring force in this example results in a non-constant acceleration model. This is confirmed by the velocity values obtained in column 4 of the table of results. Clearly the velocity is not changing at a uniform rate.

5.7 | Further discussion on examples

The work–energy procedure is most effective when applied to the solution of problems where forces are described as functions of their position rather than time or velocity. The method involves the calculation of the kinetic and strain energies of a system together with the mechanical work done. In the examples in this chapter we have assumed that we need only account for mechanical energy. This is reasonable if the work done does not generate significant amounts of other types of energy such as heat. Thermodynamics is concerned with heat and its conversion into mechanical energy. Thermodynamic analysis is beyond the scope of this text.

Work and energy can go into or come out of a system depending upon whether the generating force acts in the direction of motion of the system or opposes it. However, it should be remembered that work and energy are scalar quantities and as such have no associated direction.

5.8 | Problems

5.1 A 30 kg crate slides 3 m down a 25° slope. If the coefficient of friction between the crate and the slope is 0.15, determine the work done on the crate.

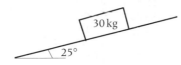

5.2 A car is travelling at a speed of 30 m s^{-1}. The car has a mass of 800 kg and four wheels each with a diameter of 0.58 m, thickness 0.15 m and mass 10 kg. Calculate the kinetic energy.

5.3 A 25 kg crate is pulled down a 20° slope by a 250 N force parallel to the slope. The coefficient of friction between the crate and the sloping surface is 0.25. At an instant the speed of the crate is found to be 1 m s^{-1}. Determine the distance that the crate has to travel down the slope before its speed becomes 6 m s^{-1}.

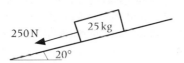

5.4 A spring has an unstretched length of 0.3 m and a stiffness of 3 kN m^{-1}.
    (a) Calculate the work needed to stretch the spring to a length of 0.32 m.
    (b) Determine the additional work to stretch the spring from 0.32 m to 0.34 m.

5.5 A truck of unknown mass is travelling at 1.6 m s^{-1}. The truck is fitted with a buffer spring with a stiffness 795 kN m^{-1}. The truck is brought to an immediate stop on collision with a solid block, resulting in the spring being compressed by 375 mm. Determine the mass of the truck.

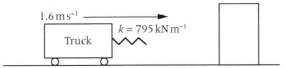

5.6 A 0.3 m diameter cylinder with a mass of 6 kg rolls 5 m down a 10° slope from a standing start. Determine the minimum value for the coefficient of friction between the two surfaces to ensure that rolling rather than sliding takes place, and calculate the final values of the translational and angular velocities of the cylinder.

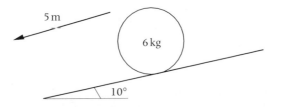

5.7   The system shown below is initially at rest. Block A has a mass of 10 kg and wheel B has a mass of 60 kg. The spring, which is initially compressed by 0.2 m, has a stiffness of 600 N m^{-1}. Determine the velocity of block A after it has been raised by 0.3 m.

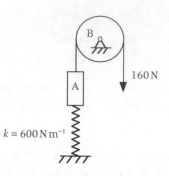

5.8   In the system shown below, the wheel has a mass of 15 kg and a diameter of 0.2 m. The spring has a stiffness of 150 N m^{-1}. The wheel is rolled to the left until the spring is compressed by 0.25 m and then released. Setting $x = 0$ in the release position, determine the angular velocity of the wheel during the subsequent motion for $0 \leq x \leq 0.5$ m in steps of 0.05 m.

CHAPTER 6

Impulse–
Momentum
Methods

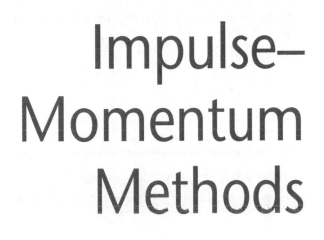

The work–energy method was introduced in Chapter 5. An alternative approach is to apply the impulse–momentum equation as described in this chapter. As in the previous chapter, we will first consider the two concepts separately before going on to use the link between them to solve a range of dynamics problems.

The following concepts are introduced in this chapter:

- *Impulse*
- *Momentum*

- *Impact*

6.1 | Linear momentum

The linear momentum of an object in translation is the product of its mass and velocity. Momentum is a vector and acts in the same direction as the object's velocity.

$$G = mv$$ (6.1)

where m = mass (kg)

v = velocity ($m\,s^{-1}$)

G = linear momentum (Ns) (since $N = kg\,m\,s^{-2}$).

EXAMPLE 6.1

Calculate the linear momentum of the following objects for the conditions described:

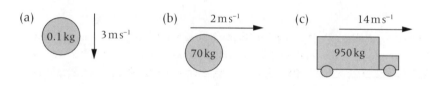

(a) 0.1 kg $3\,m\,s^{-1}$

(b) $2\,m\,s^{-1}$ 70 kg

(c) $14\,m\,s^{-1}$ 950 kg

Solution

Since momentum is a vector it is important to state direction in the answers.

(a)
$$G = mv$$
$$= (0.1)(3)$$
$$= 0.3\,Ns\downarrow$$

(b)
$$G = mv$$
$$= (70)(2)$$
$$= 140\,Ns\rightarrow$$

(c)
$$G = mv$$
$$= (950)(14)$$
$$= 13.3\,kNs\rightarrow$$

EXAMPLE 6.2

Calculate the total linear momentum of the object shown for the conditions described and the components of momentum in the X and Y directions.

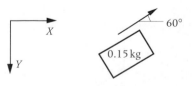

X

Y

60°

0.15 kg

Solution

(a) Total linear momentum is the product of mass and velocity and acts in the direction of the velocity.

$$G = mv$$
$$= (0.15)(3)$$
$$= 0.45\,\text{Ns}\,\angle\,60°$$

(b) The components in the X and Y directions can be calculated separately. Positive and negative signs are associated with the directions of the chosen axes.

$$G_X = mv_X$$
$$= (0.15)(3\cos 60°)$$
$$= 0.225\,\text{Ns}\rightarrow$$

$$G_Y = mv_Y$$
$$= (0.15)(-3\sin 60°)$$
$$= -0.390\,\text{Ns}\downarrow$$
$$= 0.390\,\text{Ns}\uparrow$$

6.2 | Impulse

The impulse of a force is the product of the magnitude of the force and the time interval over which it acts. Like momentum, impulse is a vector and it acts in the direction of the applied force.

$$J = \int F \cdot dt$$

For a constant force:

$$\boxed{J = Ft}$$

(6.2)

where F = force (N)
$\quad t$ = time (s)
$\quad J$ = impulse (Ns).

EXAMPLE 6.3

A mass of 2 kg drops freely for 3 seconds. Calculate the impulse on the mass.

Solution

Draw the free-body diagram to identify the forces acting.

$$J = Ft$$
$$= (2 \times 9.81)(3)$$
$$= 58.86 \text{ Ns} \downarrow$$

6.3 | Impulse–momentum method

Impulse is the means of transmission of momentum from one object to another. The final momentum in a system is simply equal to the initial momentum in the system plus the impulses applied to the system.

$$G_2 = G_1 + J \qquad\qquad (6.3)$$

where G_2 = final momentum (Ns)
G_1 = initial momentum (Ns)
J = impulse (Ns).

EXAMPLE 6.4

The impulse–momentum method is particularly useful in dynamics problems where the unknown is time. An 880 kg crate is travelling along a horizontal surface. The coefficient of friction is 0.9 and the crate is initially travelling at 30 ms^{-1}. Determine the time till the crate comes to rest under the action of the friction force.

Solution

Draw a sketch of the problem and a free-body diagram.

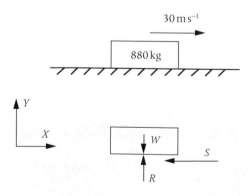

Use the translational form of Newton's second law to determine the normal reaction, R, and hence the friction force, S.

$$\sum F_Y = ma_Y = 0$$
$$R - W = 0$$
$$R = W$$
$$R = 880 \times 9.81$$
$$= 8632.8 \text{ N}$$

$$S = \mu R$$
$$= (0.9)(8632.8)$$
$$= 7769.5 \text{ N}$$

The motion is in the X-direction, so determine the initial and final momentum in the X-direction.

$$G_{1X} = mv_{1X}$$
$$= 880 \times 30$$
$$= 26\,400 \text{ N s}$$

$$G_{2X} = mv_{2X}$$
$$= 880 \times 0$$
$$= 0 \text{ N s}$$

Write an expression for the X-direction impulses applied, setting the unknown time as t. The force S is in the negative X-direction.

$$J_X = F_X t$$
$$= St$$
$$= -7769.5\,t$$

Now equate initial and final momentum using the impulse–momentum equation:

$$G_{2X} = G_{1X} + J_X$$
$$0 = 26\,400 - 7769.5t$$
$$t = \frac{26\,400}{7769.5}$$
$$t = 3.4 \text{ s}$$

The crate will take 3.4 s to come to rest under the action of the friction force.

EXAMPLE 6.5

At the instant shown in the system overleaf, mass A has a velocity of $2\,\text{m s}^{-1}\uparrow$. The coefficient of friction between the running surface and blocks A and B is 0.1. Calculate the magnitude of the force, P, such that mass A has a velocity of $5\,\text{m s}^{-1}\uparrow$ after 15 seconds.

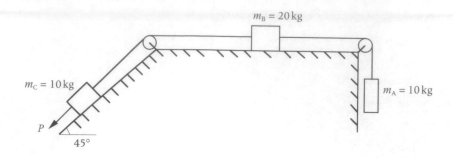

Solution

First draw the free-body diagram of the complete system apart from the running surface. The definition of the direction X has to be consistent throughout the movement of the elements which make up the complete system.

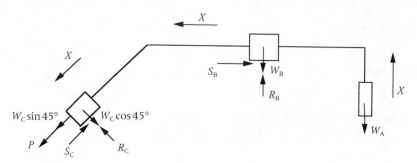

Determine the initial and final momentum of the complete system in the X-direction.

$$G_{1X} = m_A v_{1X} + m_B v_{1X} + m_C v_{1X}$$
$$= (10 \times 2) + (20 \times 2) + (10 \times 2)$$
$$= 80 \, \text{N s}$$

$$G_{2X} = m_A v_{2X} + m_B v_{2X} + m_C v_{2X}$$
$$= (10 \times 5) + (20 \times 5) + (10 \times 5)$$
$$= 200 \, \text{N s}$$

To determine the X-direction impulses we need first to calculate the friction forces S_C and S_B. As usual we do this by calculating the reaction forces and then multiplying these by the coefficient of friction. Consider masses B and C separately for this.

For mass C:

$$\sum F_Y = 0$$
$$R_C - W_C \cos 45° = 0$$
$$R_C = (10 \times 9.81) \cos 45°$$
$$= 69.367 \, \text{N}$$
$$F_C = \mu R_C$$
$$= 0.1 \times 69.367$$
$$= 6.937 \, \text{N}$$

For mass B:

$$\sum F_Y = 0$$
$$R_B - W_B = 0$$
$$R_B = 10 \times 9.81$$
$$= 98.1 \, \text{N}$$
$$F_B = \mu R_B$$
$$= 0.1 \times 98.1$$
$$= 9.81 \, \text{N}$$

Now determine the X-direction impulses for the complete system.

$$J_X = F_X t$$
$$= (P + W_C \sin 45° - F_C - F_B - W_A)t$$
$$= (P + (10 \times 9.81 \sin 45°) - 6.937 - 9.81 - (10 \times 9.81))15$$
$$- (15P - 682.197) \, \text{N s}$$

Now use the impulse–momentum equation to equate initial and final momentum. Rearrange the resulting equation to calculate the unknown, which in this case is the force P.

$$G_{2X} = G_{1X} + J_X$$
$$200 = 80 + (15P - 682.197)$$
$$P = \frac{200 - 80 + 682.197}{15}$$
$$= 53.480 \, \text{N}$$

A force, P, of $53.480 \, \text{N}$ is required to accelerate the system from $2 \, \text{ms}^{-1}$ to $5 \, \text{ms}^{-1}$ after $15 \, \text{s}$.

6.4 | Collision problems

During a collision the impact period is very short and hence any external forces have negligible impulse, so $J = 0$ and therefore in these special cases $G_2 = G_1$. This means that the momentum content of the system in a collision is constant in any given direction.

For collisions involving two masses, A and B, where we denote the velocities before and after impact by u and v respectively, we therefore have the following equation:

$$m_A u_A + m_B u_B = m_A v_A + m_B v_B \qquad (6.4)$$

Usually m_A, m_B, u_A and u_B will be known and we will be required to determine the velocities v_A and v_B after the collision. Since there are two unknowns, this equation alone is insufficient to produce an answer. We need to know a bit more about the collision. Consider the difference between a collision between two lumps of putty and two billiard balls. In the first case virtually all the energy in the collision is lost in deforming the two bodies, while in the second case virtually no energy is lost. For practical problems we need an estimate of the energy lost in the collision that we are considering. This can be provided by the coefficient of restitution, e, defined as the ratio of the separation and approach speeds.

$$e = \frac{v_B - v_A}{u_A - u_B} \qquad (6.5)$$

The coefficient of restitution varies between 0, in the case of the putty collision, to 1, for the billiard ball collision. Other typical values are 0.95 for glass, 0.7 for steel, 0.5 for wood and 0.15 for lead. Hence if we know the materials involved in the collision then we can estimate the coefficient of restitution and use the above relationship to generate a second equation in v_A and v_B. Solving the two equations simultaneously allows us to calculate the velocities resulting after the collision.

EXAMPLE 6.6

Two objects, A and B, with velocities and masses as shown below, collide. The coefficient of restitution is 0.2. Determine the velocities of the objects after the collision.

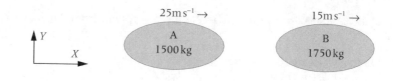

Solution

Equate the momentum in the X-direction, before and after the collision. Both velocities in this example are in the positive X-direction.

$$m_A u_{AX} + m_B u_{BX} = m_A v_{AX} + m_B v_{BX}$$
$$(1500)(25) + (1750)(15) = 1500 v_{AX} + 1750 v_{BX}$$
$$1500 v_{AX} + 1750 v_{BX} = 63\,750 \qquad \text{(eqn 1)}$$

Develop a second equation using the coefficient of restitution relationship.

$$e = \frac{v_B - v_A}{u_A - u_B}$$

$$0.2 = \frac{v_{BX} - v_{AX}}{25 - 15}$$

$$v_{BX} - v_{AX} = 2$$

$$v_{BX} = v_{AX} + 2 \qquad \text{(eqn 2)}$$

Now substitute for v_{BX} in eqn (1):

$$1500 v_{AX} + 1750(v_{AX} + 2) = 63\,750$$
$$3250 v_{AX} = 60\,250$$
$$v_{AX} = 18.54 \text{ m s}^{-1}$$

The implied positive result means that the velocity, v_{AX}, is in the positive X-direction. Now back-substitute into eqn (2) to determine v_{BX}:

$v_{BX} - v_{AX} + 2$
$v_{BX} = 18.54 + 2$
$v_{BX} = 20.54\,\mathrm{m\,s^{-1}}$

Again a positive answer, indicating that the velocity, v_{BX}, is also in the positive X-direction. Hence after the collision V_A is $18.54\,\mathrm{m\,s^{-1}} \rightarrow$ and V_B is $20.54\,\mathrm{m\,s^{-1}} \rightarrow$.

EXAMPLE 6.7

In this example the same two objects as in Example 6.6 again collide, only this time object B is travelling with a velocity of $15\,\mathrm{m\,s^{-1}}$ to the left. Again calculate the velocity of the two objects after the collision.

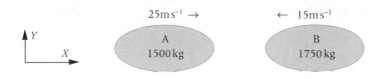

Solution

Proceed as before but take care to use $(-15\,\mathrm{m\,s^{-1}})$ as the velocity of B.

$$m_A u_{AX} + m_B u_{BX} = m_A v_{AX} + m_B v_{BX}$$
$$(1500)(25) + (1750)(-15) = 1500 v_{AX} + 1750 v_{BX}$$
$$1500 v_{AX} + 1750 v_{BX} = 11\,250 \qquad \text{(eqn 1)}$$

$$e = \frac{v_B - v_A}{u_A - u_B}$$

$$0.2 = \frac{v_{BX} - v_{AX}}{25 - (-15)}$$

$$v_{BX} - v_{AX} = 8$$
$$v_{BX} = v_{AX} + 8 \qquad \text{(eqn 2)}$$

$$1500 v_{AX} + 1750(v_{AX} + 8) = 11\,250$$
$$3250 v_{AX} = -2750$$
$$v_{AX} = -0.85\,\mathrm{m\,s^{-1}}$$

$v_{BX} = v_{AX} + 8$
$v_{BX} = -0.85 + 8$
$v_{BX} = 7.15\,\mathrm{m\,s^{-1}}$

The negative solution for v_{AX} indicates that the direction of the velocity is to the left. After the collision the velocities of A and B are $0.85\,\mathrm{m\,s^{-1}} \leftarrow$ and $7.15\,\mathrm{m\,s^{-1}} \rightarrow$ respectively.

EXAMPLE 6.8

Two objects, A and B, with velocities as shown below, collide. The coefficient of restitution is 0.3. Determine the velocities of the two objects after the collision.

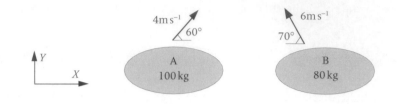

Solution

Consider the X and Y components of momentum separately. First consider the X components of momentum.

$$u_{AX} = 4\cos 60° = 2 \text{ m s}^{-1}$$
$$u_{BX} = -6\cos 70° = -2.052 \text{ m s}^{-1}$$

$$m_A u_{AX} + m_B u_{BX} = m_A v_{AX} + m_B v_{BX}$$
$$(100)(2) + (80)(-2.052) = 100 v_{AX} + 80 v_{BX}$$
$$100 v_{AX} + 80 v_{BX} = 35.84 \qquad \text{(eqn 1)}$$

$$e = \frac{v_{BX} - v_{AX}}{u_{AX} - u_{BX}}$$

$$0.3 = \frac{v_{BX} - v_{AX}}{2 - (-2.052)}$$

$$v_{BX} - v_{AX} = 1.216$$
$$v_{BX} = v_{AX} + 1.216 \qquad \text{(eqn 2)}$$

$$100 v_{AX} + 80(v_{AX} + 1.216) = 35.84$$
$$180 v_{AX} = -61.408$$
$$v_{AX} = -0.341 \text{ m s}^{-1}$$

$$v_{BX} = v_{AX} + 1.216$$
$$v_{BX} = -0.341 + 1.216$$
$$v_{BX} = -0.875 \text{ m s}^{-1}$$

The X components of the velocity of objects A and B are $0.341 \text{ m s}^{-1} \leftarrow$ and $0.875 \text{ m s}^{-1} \rightarrow$ respectively. Now consider the Y components of momentum.

$$u_{AY} = 4\sin 60° = 3.464 \text{ m s}^{-1}$$
$$u_{BY} = 6\sin 70° = 5.638 \text{ m s}^{-1}$$

$$m_A u_{AY} + m_B u_{BY} = m_A v_{AY} + m_B v_{BY}$$
$$(100)(3.464) + (80)(5.638) = 100 v_{AY} + 80 v_{BY}$$
$$100 v_{AY} + 80 v_{BY} = 797.44 \qquad \text{(eqn 3)}$$

$$e = \frac{v_{BY} - v_{AY}}{u_{AY} - u_{BY}}$$

$$0.3 = \frac{v_{BY} - v_{AY}}{3.464 - 5.638}$$

$$v_{BY} - v_{AY} = -0.652$$

$$v_{BY} = v_{AY} - 0.652 \qquad \text{(eqn 4)}$$

$$100v_{AY} + 80(v_{AY} - 0.652) = 797.44$$

$$180v_{AY} = 849.616$$

$$v_{AY} = 4.720 \text{ m s}^{-1}$$

$$v_{BY} = v_{AY} - 0.652$$
$$v_{BY} = 4.720 - 0.652$$
$$v_{BY} = 4.068 \text{ m s}^{-1}$$

The Y components of the velocity of objects A and B are $4.720 \text{ m s}^{-1} \uparrow$ and $4.068 \text{ m s}^{-1} \uparrow$ respectively. Now combine the X and Y components of velocity to determine the resultant velocities of the two objects after the collision.

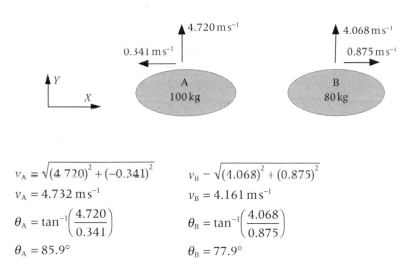

$$v_A = \sqrt{(4.720)^2 + (-0.341)^2}$$
$$v_A = 4.732 \text{ m s}^{-1}$$
$$\theta_A = \tan^{-1}\left(\frac{4.720}{0.341}\right)$$
$$\theta_A = 85.9°$$

$$v_B = \sqrt{(4.068)^2 + (0.875)^2}$$
$$v_B = 4.161 \text{ m s}^{-1}$$
$$\theta_B = \tan^{-1}\left(\frac{4.068}{0.875}\right)$$
$$\theta_B = 77.9°$$

The final outcome of the collision can hence be represented as shown below.

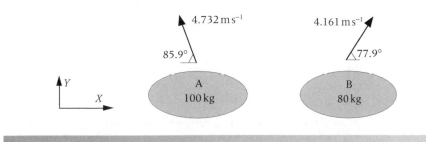

6.5 | Further discussion on examples

The impulse–momentum method is most appropriate in the solution of problems where force is described as a function of time, allowing us to calculate the velocity change that occurs, or if the velocities are known, we can determine the applied force or the time required.

In impact problems, where the interaction between two objects is of short duration, impulse–momentum can be used to assess the outcome of the impact, provided that information regarding the deformation properties of the objects involved in the collision, in the form of a coefficient of restitution, is also known.

6.6 | Problems

6.1 A 4 kg crate is moving as shown with a velocity of $2.5\,\mathrm{m\,s^{-1}}$. Calculate the linear momentum of the crate.

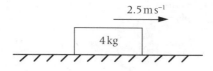

6.2 A 4 kg crate is pulled over a horizontal surface by a force of 25 N for 5 s. The coefficient of friction between the crate and the surface is 0.3. Calculate the impulse.

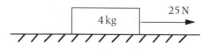

6.3 Both systems shown below are initially at rest and then released. For each case determine the velocity of the 11 kg mass.

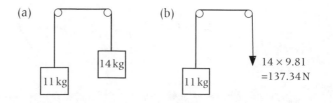

6.4 A 15 kg crate, which is initially at rest, begins to slide down a 15° slope. If the coefficient of friction between the crate and the slope is 0.2, calculate the time taken for the crate to attain a velocity of $4\,\mathrm{m\,s^{-1}}$.

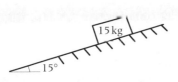

6.5 The system shown below is moving to the right with a velocity of $1\,\mathrm{m\,s}^{-1}$. The coefficient of friction between the 9 kg crate and the surface is 0.15, and the coefficient of friction between the 18 kg crate and the surface is 0.25. Determine the velocity of the system 3 s later.

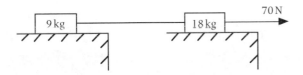

6.6 The system shown below is initially at rest. Calculate the value of the force P such that the system has a velocity of $4\,\mathrm{m\,s}^{-1}$ after 15 s. The coefficient of friction between the crates and the surface is 0.15.

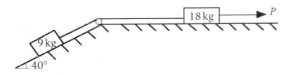

6.7 Object A, travelling at $6\,\mathrm{m\,s}^{-1}$ as shown below, strikes object B which is stationary. If the coefficient of restitution of the collision is 0.3, determine the velocities of each object immediately after the collision.

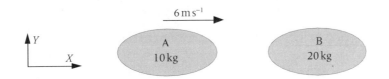

6.8 The two objects shown below are travelling with the velocities shown just before they collide. If the coefficient of restitution of the collision is 0.3, determine the velocities of each of the objects immediately after they collide.

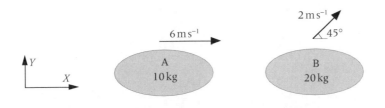

CHAPTER 7

Beam Equations

This chapter begins by introducing the concepts of bending moment and shear force. The process of writing a series of equations which describe the bending moment, M_x, on beams with respect to the distance, x, measured along the beam from the left-hand end, is explained. Simple situations are introduced first, starting with single point loads and uniformly distributed loads (UDLs). Load combination situations, which require equation limits to be specified, are then tackled, followed by the approach required for handling partial loading cases. The chapter also shows that, with the application of some simple calculus, these equations can be used to determine the position and value of maximum bending moments, points of contraflexure, slopes and deflections.

The following concepts are introduced in this chapter:

- Bending moments
- Shear forces
- Bending moment equations
- Limits of equations
- Maximum bending moments
- Contraflexure
- Slope
- Deflection
- Continuity

7.1 | Beam behaviour

Consider the loaded beam, AB, shown below.

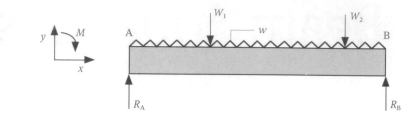

We know from equilibrium that $\Sigma M = 0$, $\Sigma F_x = 0$ and $\Sigma F_y = 0$. This is true not only for the complete system, but also holds for any subsystem. If we were to isolate a section of the beam, say AC, and draw the free-body diagram of AC, then this section would itself be in equilibrium.

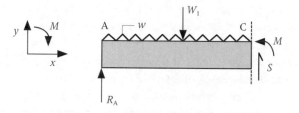

The external loading on AC creates a moment at C. Since AC is in equilibrium this must be resisted by an internal moment, M. This internal moment is referred to as the *bending moment*.

Similarly, for translational equilibrium, the beam must generate an internal force at C to balance the applied forces on AC. This force is referred to as the *shear force*. If the beam is physically unable to generate sufficient shear strength then the beam will fail by slicing apart.

7.2 | Point loads

Consider the portion of beam shown in the figure below acted upon by the point load, W. Adopting the usual convention that clockwise moments are positive and that x is a measure of position along the beam from the left-hand end, we can see that the value of the bending moment for any value of x is given by:

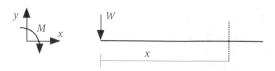

$$M_x = -Wx \qquad (7.1)$$

The negative sign occurs since W will cause an anticlockwise rotation at x.

Owing to the reaction, R, in the following example the bending moment for any value of x is given by:

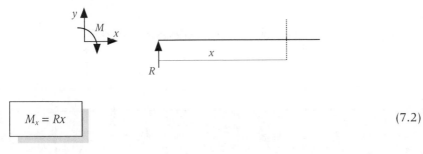

$$M_x = Rx \qquad (7.2)$$

If we now consider the case of a point load W, which is situated a distance d from the left-hand end, we need two expressions to describe the bending moment at x: one which applies for values of x less than or at most equal to d, and one which applies for values of x equal to or greater than d.

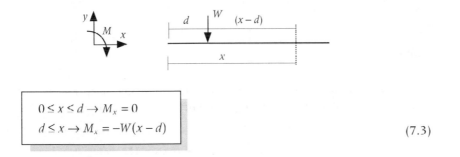

$$0 \leq x \leq d \rightarrow M_x = 0$$
$$d \leq x \rightarrow M_x = -W(x - d) \qquad (7.3)$$

7.3 | Uniformly distributed loads (UDLs)

The bending moment caused by a UDL, w $(\mathrm{N\,m^{-1}})$, which is x (m) long, is calculated using an equivalent point load, wx (N m), acting at the centre of the UDL.

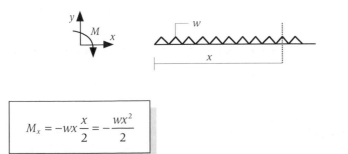

$$M_x = -wx\,\frac{x}{2} = -\frac{wx^2}{2} \qquad (7.4)$$

7.4 | Combinations of loads

If several loads act on the beam then the bending moment at x is equal to the sum of the effects that apply. This will mean that a series of equations will require to be written with appropriate limits as in the example below.

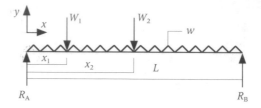

$$0 \le x \le x_1 \rightarrow M_x = R_A x - \left(\frac{wx^2}{2} \right)$$

$$x_1 \le x \le x_2 \rightarrow M_x = R_A x - \left(\frac{wx^2}{2} \right) - W_1(x - x_1)$$

$$x_2 \le x \le L \rightarrow M_x = R_A x - \left(\frac{wx^2}{2} \right) - W_1(x - x_1) - W_2(x - x_2)$$

(7.5)

EXAMPLE 7.1

Write the bending moment equations for the following beam and use these to determine the value of the bending moment at 1 m intervals along the beam.

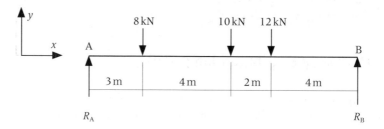

Solution

Determine the reaction force R_A and then write the equations in turn stating the appropriate limits.

$$\sum M_B = 0$$

$$13R_A - (8)(10) - (10)(6) - (12)(4) = 0$$

$$R_A = \frac{80 + 60 + 48}{13}$$

$$R_A = 14.4615 \, \text{kN}$$

$$0 \le x \le 3 \rightarrow M_x = 14.4615x$$
$$3 \le x \le 7 \rightarrow M_x = 14.4615x - 8(x-3)$$
$$7 \le x \le 9 \rightarrow M_x = 14.4615x - 8(x-3) - 10(x-7)$$
$$9 \le x \le 13 \rightarrow M_x = 14.4615x - 8(x-3) - 10(x-7) - 12(x-9)$$

The solution can be conveniently shown in tabular format:

x (m)	M_x formula	M_x (kN m)
0	$M_x = 14.4615x$	0
1		14.4615
2		28.9231
3		43.3845
3	$M_x = 14.4615x - 8(x-3)$	43.3845
4		49.8460
5		56.3075
6		62.7690
7		69.2305
7	$M_x = 14.4615x - 8(x-3) - 10(x-7)$	69.2305
8		65.6920
9		62.1535
9	$M_x = 14.4615x - 8(x-3) - 10(x-7) - 12(x-9)$	62.1535
10		46.6150
11		31.0765
12		15.5380
13		0

EXAMPLE 7.2

Write the bending moment equations for the following beam and, using these, determine the value of the bending moment at 1 m intervals along the beam.

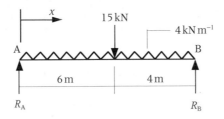

Solution

Determine the reaction force R_A and then write the equations in turn, stating the appropriate limits.

$$\sum M_B = 0$$
$$10R_A - 4(10)(5) - (15)(4) = 0$$
$$R_A = \frac{200 + 60}{10}$$
$$R_A = 26 \text{ kN}$$

$$0 \le x \le 6 \rightarrow M_x = 26x - \left(\frac{4x^2}{2}\right)$$

$$6 \le x \le 10 \rightarrow M_x = 26x - \left(\frac{4x^2}{2}\right) - 15(x - 6)$$

Setting out the solution in tabular format gives the following:

x (m)	M_x formula	M_x (kNm)
0	$M_x = 26x - 2x^2$	0
1		24
2		44
3		60
4		72
5		80
6		84
6	$M_x = 26x - 2x^2 - 15(x - 6)$	84
7		69
8		50
9		27
10		0

A visual representation of the results can be produced by plotting a bending moment diagram. Conventionally on the bending moment diagram x is plotted horizontally and M_x vertically. The bending moment diagrams for Examples 7.1 and 7.2 are shown below.

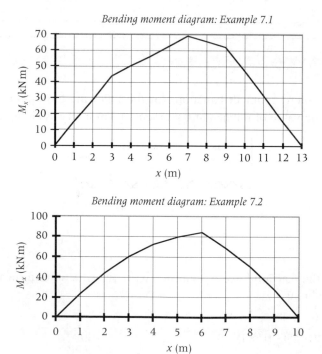

Bending moment diagram: Example 7.1

Bending moment diagram: Example 7.2

7.5 | Partial loading, late start UDLs

If the UDL acts on only part of the beam then the following approach should be adopted. There are two cases to consider. These are referred to as *late start* and *early finish*.

The beam shown below is an example of a late start loading, so called because the UDL does not contribute to the bending moment equation written between the initial limits.

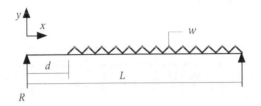

$$0 \le x \le d \rightarrow M_x = Rx$$

$$d \le x \le L \rightarrow M_x = Rx - \left[w(x-d)\left(\frac{x-d}{2}\right) \right] = Rx - \frac{w(x-d)^2}{2}$$

(7.6)

EXAMPLE 7.3

Write the bending moment equations for the following beam and use these to determine the value of the bending moment at 1 m intervals along the beam.

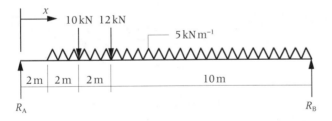

Solution

As an alternative to first determining the reactions, in this example we will write the equations directly then use the fact that the bending moment equals zero at the end of the beam, i.e. in this case at M_{16}, to deduce the reaction R_A and then complete the third column of the table.

x (m)	M_x formula	M_x (kN m)
0	$R_A x$	0
1		45.625
2		91.250
2	$R_A x - \dfrac{5(x-2)^2}{2}$	91.250
3		134.375
4		172.500
4	$R_A x - \dfrac{5(x-2)^2}{2} - 10(x-4)$	172.500
5		195.625
6		213.750
6	$R_A x - \dfrac{5(x-2)^2}{2} - 10(x-4) - 12(x-6)$	213.750
7		214.875
8		211.000
9		202.125
10		188.250
11		169.375
12		145.500
13		116.625
14		82.750
15		43.875
16		0

Note that R_A was calculated as follows:

$$M_{16} = 0$$

$$16R_A - \frac{5(16-2)^2}{2} - 10(16-4) - 12(16-6) = 0$$

$$R_A = 45.625 \, \text{kN}$$

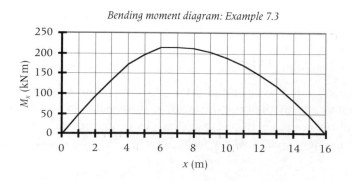

Bending moment diagram: Example 7.3

7.6 │ Partial loadings, early finish UDLs

Early finish UDL loadings are slightly trickier. The best way to treat these is to extend the loading up to the right-hand end and then to cancel out the fictitious portion of loading with an imaginary late start UDL acting upwards. The example below highlights the procedure.

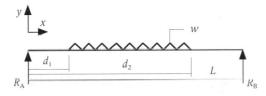

The loading can be represented by the equivalent system shown below, consisting of two early start UDLs with three bending moment equation regions. Note in particular the positive sign associated with the clockwise bending moment created by the fictional upward load w.

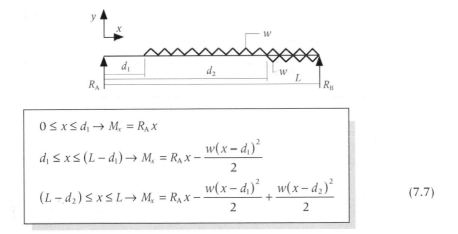

$$0 \leq x \leq d_1 \rightarrow M_x = R_A x$$

$$d_1 \leq x \leq (L - d_1) \rightarrow M_x = R_A x - \frac{w(x - d_1)^2}{2}$$

$$(L - d_2) \leq x \leq L \rightarrow M_x = R_A x - \frac{w(x - d_1)^2}{2} + \frac{w(x - d_2)^2}{2} \qquad (7.7)$$

EXAMPLE 7.4

Write the bending moment equations for the beam shown below and use these to determine the value of the bending moment at 0.5 m intervals.

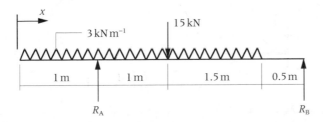

Solution

There are four intervals to consider. Treat the $3\,\mathrm{kN\,m^{-1}}$ load as if it extended all the way to the right-hand end of the beam and counteract the imaginary part with a $3\,\mathrm{kN\,m^{-1}}$ force starting late at $x = 3.5\,\mathrm{m}$ and acting upwards.

x (m)	M_x formula	M_x (kN m)
0	$-\dfrac{3x^2}{2}$	0
0.5		−0.375
1		−1.5
1	$-\dfrac{3x^2}{2} + R_A(x-1)$	−1.5
1.5		5.563
2		11.875
2	$-\dfrac{3x^2}{2} + R_A(x-1) - 15(x-2)$	11.875
2.5		9.938
3		7.250
3.5		3.813
3.5	$-\dfrac{3x^2}{2} + R_A(x-1) - 15(x-2) + \dfrac{3(x-3.5)^2}{2}$	3.813
4		0

In the table, R_A was calculated as follows:

$$M_4 = 0$$

$$-\frac{3(4)^2}{2} + R_A(4-1) - 15(4-2) + \frac{3(4-3.5)^2}{2} = 0$$

$$R_A = 17.875\,\mathrm{kN}$$

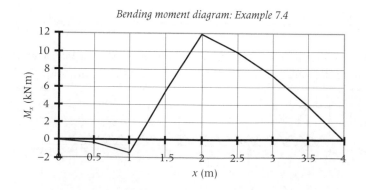

Bending moment diagram: Example 7.4

7.7 Position and value of maximum bending moments

The value of the bending moment can be calculated at regular intervals along the beams using the equations. As we have seen, it is helpful to present these in the form of a bending moment diagram in which the horizontal axis defines position along the beam corresponding to x, and with the calculated value of the bending moment, M_x, plotted on the vertical axis. When the diagram is plotted it consists of a series of M_x plots, changing from one to the other at the equation limits.

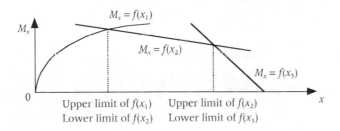

The position of any maximum values of the bending moment, including local maximums, M_{max}, can be determined from:

$$\frac{\mathrm{d}M_x}{\mathrm{d}x} = 0 \qquad\qquad (7.8)$$

The question arises: 'Which of the bending moment equations should be differentiated?' To answer this, note that the value of x calculated from the equation differentiated must lie within the equations limits, otherwise the maximum detected is imaginary. This is best illustrated by returning to the four examples just completed and determining the position and value of the maximum bending moments on these beams. We will begin by revisiting Example 7.3 rather than Examples 7.1 or 7.2, for a reason which will become evident shortly.

EXAMPLE 7.5

Determine the position and value of the maximum bending moment on the beam previously described in Example 7.3.

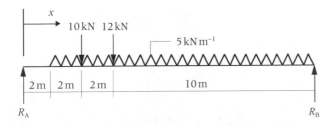

Solution

There are four bending moment equations associated with Example 7.3: M_x for $0 \le x \le 2$, M_x for $2 \le x \le 4$, M_x for $4 \le x \le 6$ and M_x for $6 \le x \le 16$. To decide which to use in order to establish the maximum bending moment, we should first examine the M_x values in the table. There are two possible candidates: $4 \le x \le 6$ or $6 \le x \le 16$. Unfortunately we cannot say for sure which is correct so we need to try one and test it. First let's assume that M_x occurs within the limits $4 \le x \le 16$:

$$M_x = 45.625x - \frac{5(x-2)^2}{2} - 10(x-4)$$

$$\frac{\mathrm{d}M_x}{\mathrm{d}x} = 45.625 - 5(x-2) - 10$$

For a maximum (or minimum) value $\dfrac{\mathrm{d}M_x}{\mathrm{d}x} = 0$ and hence:

$$45.625 - 5(x-2) - 10 = 0$$

$$x = \frac{45.625 + 10 - 10}{5}$$

$$x = 9.125\,\mathrm{m}$$

However $x = 9.125\,\mathrm{m}$ is outside the limits originally set, $4 \le x \le 6$, and therefore this maximum is imaginary. Now assume that the maximum bending moment occurs for $6 \le x \le 16$:

$$M_x = 45.625x - \frac{5(x-2)^2}{2} - 10(x-4) - 12(x-6)$$

$$\frac{\mathrm{d}M_x}{\mathrm{d}x} = 45.625 - 5(x-2) - 10 - 12$$

For a maximum (or minimum) value $\dfrac{\mathrm{d}M_x}{\mathrm{d}x} = 0$ and hence:

$$45.625 - 5(x-2) - 10 - 12 = 0$$

$$x = \frac{45.625 + 10 - 10 + 12}{5}$$

$$x = 6.725\,\mathrm{m}$$

This position is within the original limits, $6 \le x \le 16$, and it therefore corresponds to a true maximum. To determine the maximum bending moment investigate $M_{6.725}$. This is done of course by substituting 6.725 m for x using the M_x equation which applies in the limits $6 \le x \le 16$:

$$M_{\max} = M_{6.725}$$

$$M_{\max} = 45.625(6.725) - \frac{5(6.725-2)^2}{2} - 10(6.725-4) - 12(6.725-6)$$

$$M_{\max} = 223.764\,\mathrm{kN\,m}$$

The maximum bending moment on the beam has a value of 223.764 kNm and it occurs at a point 6.725 m from the left-hand end.

Now let's attempt the same procedure with Example 7.2.

EXAMPLE 7.6

Determine the position and value of the maximum bending moment on the beam previously described in Example 7.2.

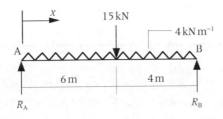

Solution

There are two bending moment equations associated with Example 7.2: M_x for $0 \leq x \leq 6$ and M_x for $6 \leq x \leq 10$. To decide which to use in order to establish the maximum bending moment, we should first examine the M_x values in the table. Again there appear to be two possibilities.

First let's assume that the maximum bending moment, M_{max}, occurs for $6 \leq x \leq 10$. In which case:

$$M_x = 26x - 2x^2 - 15(x - 6)$$

$$\frac{dM_x}{dx} = 26 - 4x - 15$$

For a maximum (or minimum) value $\dfrac{dM_x}{dx} = 0$ and hence:

$$-4x + 11 = 0$$
$$x = 2.75 \text{ m}$$

However $x = 2.75$ m is outside the limits originally set, $6 \leq x \leq 10$, and therefore this maximum is imaginary. Now assume that the maximum bending moment occurs for $0 \leq x \leq 6$:

$$M_x = 26x - 2x^2$$

$$\frac{dM_x}{dx} = 26 - 4x = 0$$

$$x = 6.5 \text{ m}$$

Again we find that the calculated position for M_{max} is outside the original limits, $0 \leq x \leq 6$, and this also must be imaginary. This then indicates that M_{max} must occur at the intersection of the two regions $0 \leq x \leq 6$ and $6 \leq x \leq 10$, that is at $x = 6\,\text{m}$. To determine M_{max} we should therefore calculate M_6 which can be done using either of the two M_x equations since $x = 6\,\text{m}$ is within the limits of both.

$$M_{max} = M_6$$
$$M_{max} = 26(6) - 2(6)^2$$
$$M_{max} = 84\ \text{kN\,m}$$

The maximum bending moment is $84\,\text{kN\,m}$ and occurs at a point $6\,\text{m}$ from the left-hand end of the beam.

7.8 | Position of the point(s) of contraflexure

In beams where one, or both, of the reactions are positioned away from the ends of the beam, such as in Example 7.4, the beam will experience sagging (downward) and hogging (upward) deflection under the applied loading.

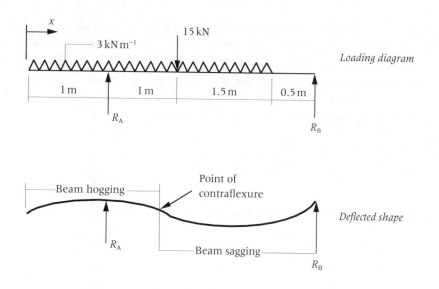

By referring to the table of bending moments produced for Example 7.4 it can be seen that positive bending moments correspond to positions on the beam with downward curvature (sagging) while negative bending moments occur where the beam has upward curvature (hogging). Where the beam shape is on the point of changing from sagging to hogging (or hogging to sagging) the bending moment is zero. Points on the beam where this occurs are known as *points of contraflexure*.

At a point of contraflexure $\boxed{M_x = 0}$ (7.9)

EXAMPLE 7.7

Determine the position of the point of contraflexure on the beam described in Example 7.4.

Solution

By referring back to the bending moment table produced for Example 7.4 we can see that the bending moment changes from being negative to positive in the range $1 \leq x \leq 2$. The point of contraflexure must therefore occur within this region of the beam and consequently the corresponding bending moment formula is the appropriate one to use.

$$M_x = 0$$

$$-\frac{3x^2}{2} + 17.875(x-1) = 0$$

$$\frac{3x^2}{2} - 17.875x + 17.875 = 0$$

This is a quadratic equation in x and hence there are two solutions for x:

$$x = \frac{17.875 \pm \sqrt{17.875^2 - (4 \times 1.5 \times 17.875)}}{2 \times 1.5}$$

$$x = 1.102\,\text{m} \quad \text{and} \quad 10.815\,\text{m}$$

Only one of the two possible answers, $x = 1.102\,$m, falls within the limits of the equation we have been using, $1 \leq x \leq 2$, and this is therefore the only solution.

The point of contraflexure occurs on the beam at a point 1.102 m from the left-hand end.

7.9 Further discussion on Examples 7.1 to 7.7

The use of bending moment equations, as described in this chapter, is a powerful routine not only for establishing the bending moments themselves, but also, as we have seen, as a means of determining the position and value of maximums and the position of points of contraflexure. With a little practice, writing the equations becomes very straightforward and routine.

It should be noted that in each example in this chapter we were dealing with a statically determinate beam. Basically this just means that the beam can be analysed using the statics methods as described in this book. For this to be the case there can only ever be two unknown reactions. Further equations can be generated by more advanced techniques for the solution of continuous structures with three or more reactions as shown below, but these are outside the scope of this text. You will encounter them later in your studies.

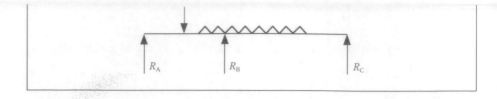

7.10 | Problems

In **7.1** to **7.3**, write the bending moment equations for the beam illustrated. Use these to calculate the value of the bending moment at 1 m intervals along the beams and sketch the bending moment diagrams.

7.1

7.2

7.3

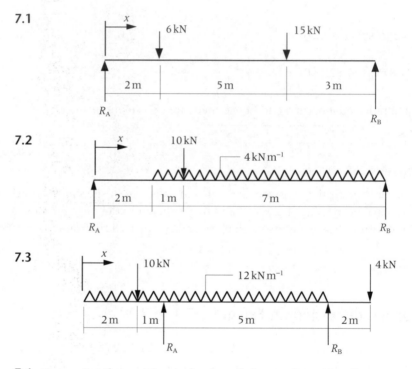

7.4 Determine the position and value of the maximum bending moment on the beam described in **7.2**.

7.5 Determine the positions and values of the maximum bending moments on the beam described in **7.3**.

7.6 Determine the position of the points of contraflexure on the beam described in **7.3**.

7.11 | The elastic curve method

Once we have written the bending moment equations and specified their limits, it is possible to generate equivalent equations to describe the shear force, slope and deflection within these same limits.

Consider a small portion of beam AB of length δx which is carrying a load of w/unit length. At A, the bending moment and shear force are M and S respectively. At B, each has changed by the small amounts δM and δS.

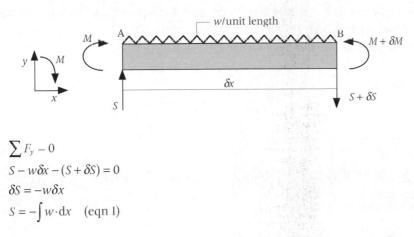

$$\sum F_y - 0$$
$$S - w\delta x - (S + \delta S) = 0$$
$$\delta S = -w\delta x$$
$$S = -\int w \cdot dx \quad \text{(eqn 1)}$$

Equation (1) shows that the shear force, S, at a point on a beam can be found by integrating the load intensity, w, with respect to the distance, x, along the beam.

$$\sum M_B = 0$$
$$M + S\delta x - (w\delta x)\frac{\delta x}{2} - (M + \delta M) = 0$$
$$S\delta x = \delta M + \frac{w\delta x^2}{2}$$

but $\delta x^2 \approx 0$
so $S\delta x = \delta M$
$$\delta M = S\delta x$$
$$M = \int S.dx \quad \text{(eqn 2)}$$

Equation (2) shows that the bending moment, M, at a point on a beam can be found by integrating the shear force, S, with respect to the distance, x, along the beam.

Now consider the curvature of this same small element of beam.

$$\tan\theta = \frac{\delta y}{\delta x}$$

for small θ, $\tan\theta = \theta$

$$\theta = \frac{\delta y}{\delta x}$$
$$\delta y = \theta\delta x$$
$$y = \int \theta.dx \quad \text{(eqn 3)}$$

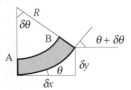

Equation (3) shows that the deflection, y, at any point on a beam can be found by integrating the slope, θ, with respect to the distance, x, along the beam. Also for small θ:

$$\delta x \approx R\delta\theta$$

$$\frac{\delta\theta}{\delta x} \approx \frac{1}{R}$$

As the beam deflects, fibres above the centroidal xx axis will shorten and those below will be stretched. The change in length of the fibres is linearly proportional to the distance of the fibre from xx.

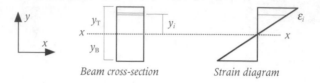

Beam cross-section   Strain diagram

At a distance y_i from the xx axis, the new length of the fibre is $(R - y_i)\delta\theta$

The change in fibre length at y_i is then $\qquad -y_i\delta\theta$

Hence the strain in the fibre at y_i is

$$\varepsilon_i = -\frac{y_i\delta\theta}{\delta x}$$

$$\varepsilon_i = -\frac{y_i}{R}$$

The stress at y_i is given by $\qquad \sigma_i = \varepsilon_i E$

$$\sigma_i = \frac{-Ey_i}{R}$$

If δA is the area of the strip of beam at y_i then

The total compressive force on the cross-section is

$$\int_0^{y_T} \sigma \cdot dA$$

$$= \int_0^{y_T} \frac{Ey}{R} \cdot dA$$

The total tensile force on the cross-section is

$$\int_0^{y_B} \sigma \cdot dA$$

$$= \int_0^{y_B} \frac{Ey}{R} \cdot dA$$

Therefore the total force on the cross-section, for constant E and R is

$$\frac{E}{R} \int_{y_B}^{y_T} y \cdot dA$$

Hence the total moment of resistance of the cross-section is

$$M = \frac{E}{R} \int_{y_B}^{y_T} y^2 \cdot dA$$

but

$$\int_{y_B}^{y_T} y^2 \cdot dA = I_A$$

hence

$$\boxed{M = \frac{EI_A}{R}}$$

$$(7.10)$$

This equation expresses the moment of resistance of a cross-section, M (Nm), in terms of the material's Young's modulus, E (Nm^{-2}), second moment of area, I_A (m^4) and the radius of curvature, R (m).

$$\frac{1}{R} = \frac{M}{EI_A}$$

$$\frac{\delta\theta}{\delta x} = \frac{M}{EI_A}$$

$$\delta\theta = \frac{1}{EI_A} \cdot M \cdot \delta x$$

$$\theta = \frac{1}{EI_A} \int M \cdot dx \quad \text{(eqn 4)}$$

Equation (4) shows that the slope, θ, at a point on a beam can be found by integrating the moment, M, with respect to the distance, x, along the beam and dividing by the stiffness, EI_A.

Taking these four equations together we have:

$$
\boxed{
\begin{aligned}
S_x &= -\int w_x \cdot dx \\
M_x &= \int S_x \, dx \\
\theta_x &= \frac{1}{EI_A} \int M_x \cdot dx \\
Y_x &= \int \theta_x dx
\end{aligned}
}
\tag{7.11}
$$

$$
\boxed{
\begin{aligned}
-w_x &= \frac{dS_x}{dx} \\
S_x &= \frac{dM_x}{dx} \\
M_x &= EI_A \frac{d\theta_x}{dx} \\
\theta_x &= \frac{dY_x}{d_x}
\end{aligned}
}
\tag{7.12}
$$

where w_x = the load intensity (Nm^{-1}) at x (m)
  S_x = the shear force (N) at x (m)
  M_x = the bending moment (Nm) at x (m)
  θ_x = the slope (rad) at x (m)
  Y_x = the deflection (m) at x (m).

The following important facts are evident:

1. Maximum bending moment(s) occur(s)   That is when $\dfrac{dM_x}{dx} = 0$.

   · where $S_x = 0$.

2. Maximum slope(s) occur(s) where $M_x = 0$.   That is when $\dfrac{d\theta_x}{dx} = 0$.

3. Maximum deflection(s) occur(s) where $\theta_x = 0$.   That is when $\dfrac{dY_x}{dx} = 0$.

4. Points of contraflexure occur where $M_x = 0$.

EXAMPLE 7.8

The 20 m span beam shown below has a Young's modulus, E, of $15\,\text{kN}\,\text{mm}^{-2}$. Neglecting the self-weight of the beam, determine the shear force, bending moment, slope and deflection equations. Calculate the deflection of the beam at 1 m intervals. State the position and value of the maximum deflection on the beam.

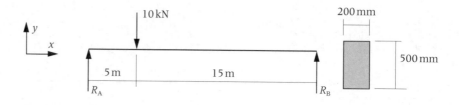

Solution

Determine the reaction force, R_A, and then write the bending moment equations. The shear force equations can be generated by differentiating the bending moment equations.

$$\sum M_{R_B} = 0$$
$$20R_A - (10 \times 15) = 0$$
$$R_A = 7.5 \text{ kN}$$

$0 \le x \le 5$

$M_x = 7.5x$

$S_x = \dfrac{dM_x}{dx}$

$S_x = 7.5$

$5 \le x \le 20$

$M_x = 7.5x - 10(x - 5)$

$S_x = \dfrac{dM_x}{dx}$

$S_x = 7.5 - 10$

Starting again with the bending moment equations, develop the slope and deflection equations for the two regions independently. Notice that this will introduce four constants of integration.

$0 \leq x \leq 5$

$M_x = 7.5x$

$$\theta_x = \frac{1}{EI_A} \int M_x . dx$$

$$\theta_x = \frac{1}{EI_A} \int (7.5x) . dx$$

$$\theta_x = \frac{1}{EI_A} \int (3.75x^2 + A)$$

(eqn 1)

$$Y_x = \int \theta_x . dx$$

$$Y_x = \frac{1}{EI_A} \int (3.75x^2 + A) . dx$$

$$Y_x = \frac{1}{EI_A} (1.25x^3 + Ax + B)$$

(eqn 2)

$5 \leq x \leq 20$

$M_x = 7.5x - 10(x - 5)$

$$\theta_x = \frac{1}{EI_A} \int M_x . dx$$

$$\theta_x = \frac{1}{EI_A} \int (7.5x - 10(x - 5)) . dx$$

$$\theta_x = \frac{1}{EI_A} \int \left(3.75x^2 - 5(x - 5)^2 + C \right)$$

(eqn 3)

$$Y_x = \int \theta_x . dx$$

$$Y_x = \frac{1}{EI_A} \int \left(3.75x^2 - 5(x - 5)^2 + C \right) . dx$$

$$Y_x = \frac{1}{EI_A} \left(1.25x^3 - \frac{5(x - 5)^3}{3} + Cx + D \right)$$

(eqn 4)

In order to eliminate the four constants of integration we require four known solutions. The first two are the end conditions that the deflections at the reactions are zero. We can therefore substitute $Y = 0$ at $x = 0$ into eqn (2) and $Y = 0$ at $x = 20$ into eqn (4).

At $x = 0$ $Y_x = 0$

$$0 = \frac{1}{EI_A} \left(1.25(0)^3 + A(0) + B \right)$$

$B = 0$ (eqn 5)

At $x = 20$ $Y_x = 0$

$$0 = \frac{1}{EI_A} \left(1.25(20)^3 - \frac{5(20 - 5)^3}{3} + C(20) + D \right)$$

$4375 + 20C + D = 0$ (eqn 6)

We can generate two more known conditions from consideration of the continuity of the beam. These are that at $x = 5$ the deflections obtained from eqns (2) and (4) must be the same, and also at $x = 5$ the slopes obtained from eqns (1) and (3) must be the same. Taking the continuity of slope first:

At $x = 5$

$$\frac{1}{EI_A} (3.75x^2 + A) = \frac{1}{EI_A} \left(3.75x^2 - 5(x - 5)^2 + C \right)$$

$$3.75(5)^2 + A = 3.75(5)^2 - 5(5 - 5) + C$$

$$A = C \quad \text{(eqn 7)}$$

For continuity of deflection we have:

At $x = 5$

$$\frac{1}{EI_A} (1.25x^3 + Ax + B) = \frac{1}{EI_A} \left(1.25x^3 - \frac{5(x - 5)^3}{3} + Cx + D \right)$$

$$1.25(5)^3 + 5A + B = 1.25(5)^3 - \frac{5(5-5)^3}{3} + 5C + D$$

$$5A + B = 5C + D$$

$$5A + B = 5A + D$$

$$B = D \qquad\qquad \text{(eqn 8)}$$

There is now sufficient information to determine the four constants of integration by solving eqns (5), (6), (7) and (8) simultaneously.

$$B = 0 \quad \text{(eqn 5)}$$

$$4375 + 20C + D = 0 \quad \text{(eqn 6)}$$

$$A = C \quad \text{(eqn 7)}$$

$$B = D \quad \text{(eqn 8)}$$

Note that eqns (7) and (8) follow directly from our consideration of continuity and will always hold. Consequently there is in fact no need to develop separate constants of integration for each equation range. Substituting into eqn (6) gives the solution for C, and subsequently A.

$$4375 + 20C + D = 0$$

$$4375 + 20C + 0 = 0$$

$$C = -218.75$$

Summarising $A = C = -218.75$ $B = D = 0$

We can now rewrite the complete set of shear force, bending moment, slope and deflection equations for both regions, free of any constants of integration:

$0 \le x \le 5$ $5 \le x \le 20$

$$S_x = 7.5 \qquad\qquad\qquad\qquad S_x = 7.5 - 10$$

$$M_x = 7.5x \qquad\qquad\qquad\qquad M_x = 7.5x - 10(x - 5)$$

$$\theta_x = \frac{1}{EI_A}(3.75x^2 - 218.75) \qquad \theta_x = \frac{1}{EI_A}\left(3.75x^2 - 5(x-5)^2 - 218.75\right)$$

$$Y_x = \frac{1}{EI_A}(1.25x^3 - 218.75x) \qquad Y_x = \frac{1}{EI_A}\left(1.25x^3 - \frac{5(x-5)^3}{3} - 218.75x\right)$$

The value of the stiffness term, $\frac{1}{EI}$, can be evaluated from the given beam data.

Care needs to be taken to ensure consistency of units. The equations that we have developed arose from loads in kN and distances in m, so we must calculate the stiffness term in the same way.

$$I_A = \frac{ab^3}{12}$$

$$= \frac{0.2 \times 0.5^3}{12} \qquad\qquad \frac{1}{EI_A} = \frac{1}{(15 \times 10^6)(0.00208)}$$

$$= 0.002083\,\text{m}^4 \qquad\qquad\qquad = 3.2005 \times 10^{-5}\,\text{m}^2\,\text{kN}^{-1}$$

Using this value in the equations will generate values for beam slope, θ, in rad, and deflection, Y, in m. By substituting for x in the shear force, bending moment, slope and deflection equations, the following table of results can be obtained:

x (m)	Y (mm)	θ (radians)	M (kN m)	S (kN)
0	0	−0.00700	0	7.5
1	−6.96	−0.00688	7.5	7.5
2	−13.68	−0.00652	15	7.5
3	−19.92	−0.00592	22.5	7.5
4	−25.44	−0.00524	30	7.5
5	−30.00	−0.00400	37.5	7.5 −2.5
6	−33.42	−0.00284	35	−2.5
7	−35.71	−0.00176	32.5	−2.5
8	−36.97	−0.00076	30	−2.5
9	−37.26	0.00016	27.5	−2.5
10	−36.67	0.00100	25	−2.5
11	−35.29	0.00176	22.5	−2.5
12	−33.18	0.00244	20	−2.5
13	−30.43	0.00304	17.5	−2.5
14	−27.12	0.00356	15	−2.5
15	−23.34	0.00400	12.5	−2.5
16	−19.15	0.00436	10	−2.5
17	−14.64	0.00464	7.5	−2.5
18	−9.89	0.00484	5	−2.5
19	−4.99	0.00496	2.5	−2.5
20	0	0.00500	0	−2.5

The deflection can be shown on a diagram. The horizontal axis corresponds to positions along the beam while the vertical axis plots deflection.

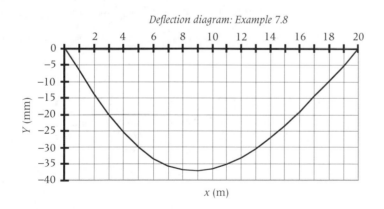

Deflection diagram: Example 7.8

From the table of deflection and slope values it is clear that the maximum deflection (and consequently zero slope) occurs somewhere near $x = 9$ m. To determine the exact position, use the slope equation for the region $5 \leq x \leq 20$ and set it equal to zero.

$$\vartheta_x = \frac{1}{EI_A}\left(3.75x^2 - 5(x-5)^2 - 218.75\right)$$

$$0 = \frac{1}{EI_A}\left(3.75x^2 - 5(x-5)^2 - 218.75\right)$$

$$3.75x^2 - 5x^2 + 50x - 125 - 218.75 = 0$$

$$1.25x^2 - 50x + 343.75 = 0$$

$$x = \frac{50 \pm \sqrt{50^2 - 4(1.25)(343.75)}}{(2)(1.25)}$$

$$x = 8.82\,\text{m}$$

Maximum deflection thus occurs at $x = 8.82$ m, obtained using the negative sign. The alternative answer, using the positive sign, gives an answer outside the equation limits and can thus be discarded. The value of the maximum deflection can now be found, determining the deflection when $x = 8.82$ m:

$$Y_{max} = Y_{8.82} = \frac{1}{EI_A}\left(1.25(8.82)^3 - \frac{5(8.82-5)^3}{3} - 218.75(8.82)\right)$$

$$= 0.03727\,\text{m}$$

$$= 37.27\,\text{mm}$$

The maximum deflection is 37.27 mm and occurs at 8.82 m from the left-hand end of the beam.

EXAMPLE 7.9

The 6 m span beam shown has a Young's modulus, E, of $210\,\text{kN}\,\text{mm}^{-2}$. In addition to the 15 kN point load, the beam carries a uniformly distributed load of $5\,\text{kN}\,\text{m}^{-1}$ which includes its self-wight. Determine the shear force, bending moment, slope and deflection of the beam at 0.25 m intervals.

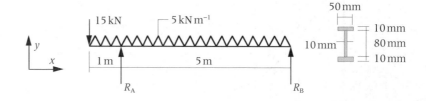

Solution

Write a separate set of equations for each region. However, as we saw in Example 7.8, owing to continuity, the same constants of integration apply in both regions.

$$\sum M_{R_B} = 0$$

$$5R_A - (15 \times 6) - (5 \times 6)3 = 0$$

$$R_A = \frac{90 + 90}{5}$$

$$R_A = 36\,\text{kN}$$

$0 \le x \le 1$

$$M_x = -15x - \frac{5x^2}{2}$$

$$S_x = \frac{dM_x}{dx}$$

$$S_x = -15 - 5x$$

$$\theta_x = \frac{1}{EI_A} \int M_x \,.dx$$

$$\theta_x = \frac{1}{EI_A} \int \left(-15x - \frac{5x^2}{2} \right) .dx$$

$$\vartheta_x = \frac{1}{EI_A} \left(-\frac{15x^2}{2} - \frac{5x^3}{6} + A \right)$$

$$Y_x = \int \theta_x \,.dx$$

$$Y_x = \frac{1}{EI_A} \int \left(-\frac{15x^2}{2} - \frac{5x^3}{6} + A \right) .dx$$

$$Y_x = \frac{1}{EI_A} \left(-\frac{15x^3}{6} - \frac{5x^4}{24} + Ax + B \right)$$

$1 \le x \le 6$

$$M_x = -15x - \frac{5x^2}{2} + 36(x - 1)$$

$$S_x = \frac{dM_x}{dx}$$

$$S_x = -15 - 5x + 36$$

$$\theta_x = \frac{1}{EI_A} \int M_x \,.dx$$

$$\theta_x = \frac{1}{EI_A} \int \left(-15x - \frac{5x^2}{2} + 36(x - 1) \right) .dx$$

$$\vartheta_x = \frac{1}{EI_A} \left(-\frac{15x^2}{2} - \frac{5x^3}{6} + 18(x - 1)^2 + A \right)$$

$$Y_x = \int \theta_x \,.dx$$

$$Y_x = \frac{1}{EI_A} \int \left(-\frac{15x^2}{2} - \frac{5x^3}{6} + 18(x - 1)^2 + A \right) .dx$$

$$Y_x = \frac{1}{EI_A} \left(-\frac{15x^3}{6} - \frac{5x^4}{24} + 6(x - 1)^3 + Ax + B \right)$$

Apply the boundary conditions, $Y_x = 0$ at $x = 1$ and $Y_x = 0$ at $x = 6$:

At $\quad x = 1 \quad Y_x = 0$

$$0 = \frac{1}{EI_A} \left(-\frac{15(1)^3}{6} - \frac{5(1)^4}{24} + A(1) + B \right)$$

$$A + B = \frac{65}{24} \quad \text{(eqn 1)}$$

At $\quad x = 6 \quad Y_x = 0$

$$0 = \frac{1}{EI_A} \left(-\frac{15(6)^3}{6} - \frac{5(6)^4}{24} + 6(6-1)^3 + A(6) + B \right)$$

$$6A + B = 60 \quad \text{(eqn 2)}$$

Solve eqns(1) and (2) simultaneously:

$$A + B = \frac{65}{24}$$

$$B = \frac{65}{24} - A$$

$$6A + B = 60$$

$$6A + \left(\frac{65}{24} - A \right) = 60$$

$$A = 11.458$$

$$B = \frac{65}{24} - 11.458$$

$$B = -8.750$$

Now write out the equation sets, with those for slope and deflection free from the constants of integration:

$$0 \le x \le 1$$

$$S_x = -15 - 5x$$

$$M_x = -15x - \frac{5x^2}{2}$$

$$\theta_x = \frac{1}{EI_A} \left(-\frac{15x^2}{2} - \frac{5x^3}{6} + 11.458 \right)$$

$$Y_x = \frac{1}{EI_A} \left(-\frac{15x^3}{6} - \frac{5x^4}{24} + 11.458x - 8.750 \right)$$

$$1 \le x \le 6$$

$$S_x = -15 - 5x + 36$$

$$M_x = -15x - \frac{5x^2}{2} + 36(x - 1)$$

$$\theta_x = \frac{1}{EI_A} \left(-\frac{15x^2}{2} - \frac{5x^3}{6} + 18(x - 1)^2 + 11.458 \right)$$

$$Y_x = \frac{1}{EI_A} \left(-\frac{15x^3}{6} - \frac{5x^4}{24} + 6(x - 1)^3 + 11.458x - 8.750 \right)$$

Finally calculate the beam stiffness term, $\dfrac{1}{EI_A}$, and present the results in a table.

The second moment of area of the section, I, is calculated by assuming a rectangular section, 50 mm × 100 mm, and then subtracting the effect of the two rectangular 20 mm × 80 mm areas.

$$I_A = \frac{0.05 \times 0.1^3}{12} - \frac{2(0.02 \times 0.08^3)}{12} \qquad \frac{1}{EI_A} = \frac{1}{(210 \times 10^6)(2.46 \times 10^{-6})}$$

$$= 2.46 \times 10^{-6} \text{ m}^4 \qquad\qquad\qquad = 0.001936 \text{ m}^2 \text{ kN}^{-1}$$

x (m)	Y (mm)	θ (radians)	M (kNm)	S (kN)
0	−16.94	0.02218	0	−15.00
0.25	−11.47	0.02125	−3.906	−16.25
0.5	−6.48	0.01835	−8.125	−17.50
0.75	−2.48	0.01334	−12.656	−18.75
1	0	0.00605	−17.500	−20.00 16.00
1.25	0.53	0.00148	−13.656	14.75
1.5	−0.59	−0.00722	−10.125	13.50
1.75	−2.94	−0.01133	−6.906	12.25
2	−6.13	−0.01396	−4.000	11.00
2.25	−9.81	−0.01525	−1.406	9.75
2.5	−13.66	−0.01537	0.875	8.50
2.75	−17.41	−0.01446	2.844	7.25
3	−20.81	−0.01267	4.500	6.00
3.25	−23.68	−0.01015	5.844	4.75
3.5	−25.84	−0.00706	6.875	3.50
3.75	−27.17	−0.00354	7.594	2.25
4	−27.59	0.00024	8.000	1.00
4.25	−27.04	0.00415	8.094	−0.25
4.5	−25.52	0.00803	7.875	−1.50
4.75	−23.05	0.01172	7.344	−2.75
5	−19.69	0.01508	6.500	−4.00
5.25	−15.54	0.01796	5.344	−5.25
5.5	−10.76	0.02021	3.875	−6.50
5.75	−5.51	0.02166	2.094	−7.75
6	0	0.02218	0	−9.00

Note from the second column of the table that there is a slight uplift in the region of the beam just to the right of reaction R_A.

If we were to attempt to find the position and value of the maximum deflection in this example by setting the slope equation equal to zero, we would be faced with solving a cubic equation. This is possible but is beyond the scope of this text and therefore we will content ourselves with an approximate maximum deflection value of 27.6 mm obtained from the table. Close inspection of the deflection column of the table shows that this is a good approximation and well within the accuracy required in any practical application.

From the table we can produce a complete set of diagrams for Example 7.9:

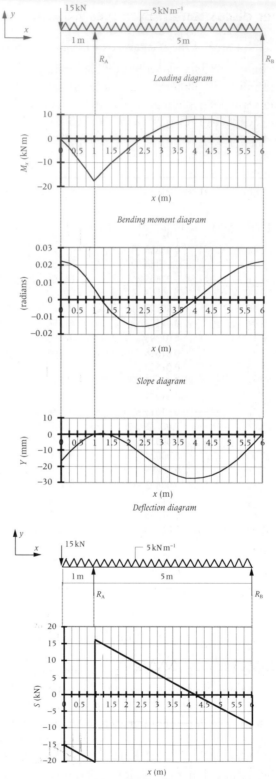

Loading diagram

Bending moment diagram

Slope diagram

Deflection diagram

Shear force diagram

7.12 Further discussion on Examples 7.8 and 7.9

In these two examples we started with the basic bending moment equations and developed the corresponding shear force, slope and deflection equations from them. The integration procedure introduced further constants of integration at each step. These constants were eliminated by investigating known solutions at specific positions on the beam. Don't be tempted to multiply out the individual terms which arise separately from each of the loadings or combine them at any intermediate step in the integration processes. Each term must be kept distinct all the way through the process. The final results are best shown as a series of diagrams, as these give a comprehensive picture of the effects at all points on the complete system.

7.13 Problems (continued)

7.7 The 8m span beam shown below has a Young's modulus, E, of 18kN mm^{-2}. Neglecting the self-weight of the beam, determine the shear force, bending moment, slope and deflection equations. Calculate the deflection of the beam at 0.5m intervals. State the position and value of the maximum deflection on the beam.

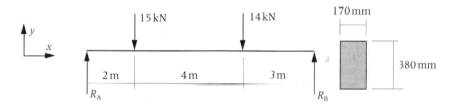

7.8 The 10m span beam shown has a Young's modulus, E, of 20kN mm^{-2}. The beam carries a uniformly distributed load of 7kN m^{-1} which includes its self-weight. Determine the shear force, bending moment, slope and deflection of the beam at 1m intervals.

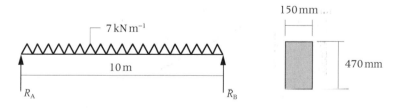

7.9 The 6m span beam shown has a Young's modulus, E, of 220kN mm^{-2}. In addition to the point loads, the beam carries a uniformly distributed load of 4kN m^{-1}

which includes its self-weight. Determine the shear force, bending moment, slope and deflection of the beam at 0.5 m intervals.

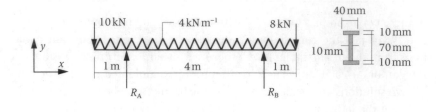

CHAPTER 8

Vibration Theory

This chapter introduces the basic concepts of vibration theory applied to one degree of freedom, undamped systems, i.e. systems which can be modelled by a single mass and a single spring. The chapter initially considers the natural vibration of this type of system. Once these ideas are established, the response of one degree of freedom systems to forced vibrations is developed and the consequences of resonance are highlighted.

The following concepts are introduced in this chapter:

- *Natural frequency*
- *Amplitude*
- *Period*

- *Resonance*
- *Transmissibility*
- *Magnification ratio*

8.1 | Natural vibrations: one degree of freedom, undamped systems

Figure 8.1(a) shows a system consisting of a single spring of stiffness k ($\mathrm{N\,m^{-1}}$) and mass m (kg) in its initial equilibrium position. The mass causes the spring to extend by an amount e (m). Figure 8.1(b) shows the corresponding free-body diagram of the mass.

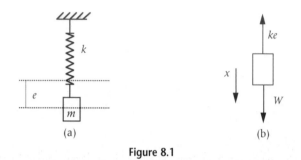

(a) (b)

Figure 8.1

The system is now displaced by a further amount, x (m), and then released. Figure 8.2(a) shows the system once released and Figure 8.2(b) shows the new free-body diagram of the mass.

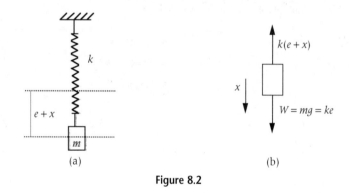

(a) (b)

Figure 8.2

From Newton's second law and referring to Figure 8.2(b) we have:

$$\sum F = m\ddot{x}$$
$$ke - k(e + x) = m\ddot{x}$$
$$m\ddot{x} + kx = 0$$

This equation is the differential equation of motion of a one degree of freedom, undamped system. To solve this second-order differential equation, first determine the roots of the auxiliary equation. If you are unfamiliar with the methods for the solution of differential equations you may well want to skip the following explanation, accept the solution and move on directly to eqn (8.1).

$$mn^2 + k = 0$$

$$n = \frac{0 \pm \sqrt{0 - 4mk}}{2m}$$

$$n = \pm \frac{2\sqrt{-mk}}{2m}$$

$$n = \pm(m)^{\frac{1}{2}}(k)^{\frac{1}{2}}(m)^{-1}(i)$$

$$n = \pm m^{-\frac{1}{2}}k^{\frac{1}{2}}i$$

$$n = \pm(i)\left(\sqrt{\frac{k}{m}}\right)$$

$$n = \alpha \pm i\beta$$

$$x = Ce^{\alpha t}\sin(\beta t + \varepsilon)$$

Let $\sqrt{\dfrac{k}{m}} = \omega$. Hence since $\alpha = 0$ and $\beta = \omega$, the solution of the second-order differential equation is:

$$x = C\sin(\omega t + \varepsilon) \tag{8.1}$$

where:

$$\omega = \sqrt{\frac{k}{m}} \tag{8.2}$$

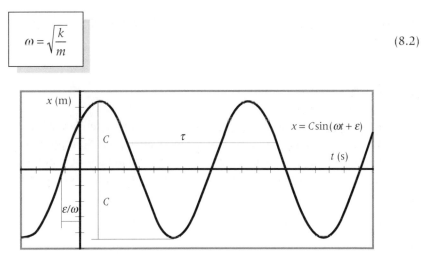

Note that the equation for the displacement, x (m), is sinusoidal and a function of time t (s). The function has a period τ (s) with repeating maximum and minimum values of x given by C, the amplitude. The natural frequency of the system, ω, is measured in rad s^{-1}. We could specify the natural frequency in cycles per second, f, or Hertz (Hz) where:

$$f = \frac{\omega}{2\pi} \tag{8.3}$$

and

$$\tau = \frac{1}{f}$$ (8.4)

It is useful to convince yourself that the equation really is a solution, particularly if you decided to skip the previous section where it was developed. Starting with our equation for displacement, x, we can develop the resulting equations for velocity and displacement:

$$x = C \sin(\omega t + \varepsilon)$$
$$\dot{x} = \omega C \cos(\omega t + \varepsilon)$$
$$\ddot{x} = -\omega^2 C \sin(\omega t + \varepsilon)$$

Now substitute back into the left-hand side (LHS) of eqn (1):

$$\begin{aligned}
\text{LHS} &= m\ddot{x} + kx \\
&= m(-\omega^2 C \sin(\omega t + \varepsilon)) + k(C \sin(\omega t + \varepsilon)) \\
&= m(-\omega^2 C \sin(\omega t + \varepsilon)) + \omega^2 m(C \sin(\omega t + \varepsilon)) \\
&= 0 \\
&= \text{RHS}
\end{aligned}$$

Thus we have proved that $x = C\sin(\omega t + \varepsilon)$ is indeed a solution to the second-order differential equation.

EXAMPLE 8.1

A system modelled as a spring of stiffness $25\,\mathrm{N\,m^{-1}}$ and a single mass of $0.5\,\mathrm{kg}$ is released from a point $400\,\mathrm{mm}$ below its steady-state equilibrium position with an upward speed of $3\,\mathrm{m\,s^{-1}}$. Determine the natural frequency of vibration of the system and the period. State the amplitude and sketch the subsequent vibration of the system.

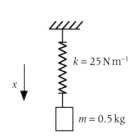

Solution

First determine the natural vibration in $\mathrm{rad\,s^{-1}}$; convert the answer into Hz and subsequently determine the period. Take care with all calculations to ensure consistency of units. Always enter the mass, m, in kg, and the stiffness, k, in $\mathrm{N\,m^{-1}}$.

$$\omega = \sqrt{\frac{k}{m}} \qquad\qquad f = \frac{\omega}{2\pi} \qquad\qquad \tau = \frac{1}{f}$$
$$= \sqrt{\frac{25}{0.5}} \qquad\qquad = \frac{7.071}{2\pi} \qquad\qquad = \frac{1}{1.125}$$
$$= 7.071\,\mathrm{rad\,s^{-1}} \qquad = 1.125\,\mathrm{Hz} \qquad\quad = 0.889\,\mathrm{s}$$

The natural frequency of vibration of the system is 7.071 rad s^{-1}, 1.125 Hz (cycles per second), giving a period of 0.889 s.

We can now set up the vibration equation of the system for the given initial displacement and velocity conditions. Begin by substituting the given initial conditions into the basic displacement and velocity equations. Note that by setting x as positive downwards, the given velocity is negative.

The initial conditions are: $x = 0.4$ m when $t = 0$ s and $\dot{x} = -3$ m s^{-1} when $t = 0$ s.

$$x = C \sin(\omega t + \varepsilon) \qquad\qquad \dot{x} = \omega C \cos(\omega t + \varepsilon)$$
$$0.4 = C \sin((7.071)(0) + \varepsilon) \qquad -3 = 7.071 C \cos((7.071)(0) + \varepsilon)$$
$$0.4 = C \sin \varepsilon \quad \text{(eqn 1)} \qquad -0.424 = C \cos \varepsilon \quad \text{(eqn 2)}$$

Solve eqns (1) and (2) simultaneously to determine C and ε. Divide eqn (1) by eqn (2):

$$\frac{0.4}{-0.424} = \frac{C \sin \varepsilon}{C \cos \varepsilon}$$
$$-0.943 = \tan \varepsilon$$

So $\tan \varepsilon$ is negative. Also since the amplitude, C, is always positive, then from eqn (1), $\sin \varepsilon$ is positive and from eqn (2), $\cos \varepsilon$ is negative. Therefore ε must be a second quadrant angle. It must be an angle between 90° and 180°.

$$\varepsilon = 180° - 43.32° = 136.68° = 2.39 \text{ rad}$$

Substitute for ε in eqn (1). Take care to set your calculator into the appropriate mode before entering the trigonometric value.

$$0.4 = C \sin \varepsilon$$
$$0.4 = C \sin(2.39)$$
$$C = \frac{0.4}{0.683}$$
$$= 0.586 \text{ m}$$

We are now in a position to write the vibration equation for the system that applies for the given initial conditions.

$$x = C \sin(\omega t + \varepsilon)$$
$$x = 0.586 \sin(7.071 t + 2.39)$$

The amplitude of the resulting vibrations is 0.586 m.

A sketch of the vibration can be obtained by substituting values for time, t, in seconds into the vibration equation and calculating the resulting displacements. The results are shown below in tabular format and overleaf in graphical format.

t (s)	0	0.1	0.2	0.3	0.4	0.5	0.6	0.7	0.8	0.9	1
x (m)	0.40	0.03	−0.36	−0.57	−0.51	−0.20	0.20	0.51	0.57	0.36	−0.02

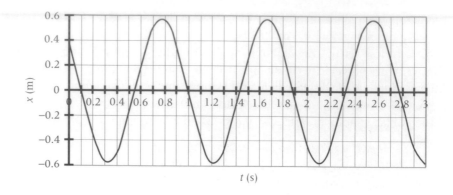

We have of course neglected the effect of damping. Our solution shows the vibration continuing undiminished with time. For a real system, damping would result in a reduction in the displacement experienced at each peak.

8.2 | Forced vibration by ground motion: one degree of freedom, undamped systems

Now consider the situation where the spring/mass system is subjected to a continuous input in the form of a sinusoidal ground motion of amplitude X_0 and frequency Ω. Figure 8.3(a) shows the details of the system and Figure 8.3(b) is the corresponding free-body diagram of the mass.

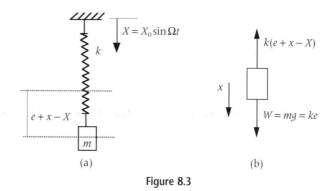

Figure 8.3

Applying Newton's second law to Figure 8.3(b) leads to:

$$\Sigma F_x = m\ddot{x}$$
$$ke - k(e + x - X) = m\ddot{x}$$
$$m\ddot{x} + kx = kX$$
$$m\ddot{x} + kx = kX_0 \sin \Omega t$$

This is the differential equation of motion of a one degree of freedom, undamped system subjected to a sinusoidal input ground motion. The solution to this equation, which we will not attempt to develop, is $x = C \sin \Omega t$.

$x = C \sin \Omega t$

$\dot{x} = \Omega C \cos \Omega t$

$\ddot{x} = -\Omega^2 C \sin \Omega t$

Substituting these into the differential equation of motion:

$$m\ddot{x} + kx = kX_0 \sin \Omega t$$

$$m(-\Omega^2 C \sin \Omega t) + k(C \sin \Omega t) = kX_0 \sin \Omega t$$

$$(-m\Omega^2 + k)C \sin \Omega t = kX_0 \sin \Omega t$$

$$(-m\Omega^2 + k)C = kX_0$$

$$C = \frac{kX_0}{(k - m\Omega^2)}$$

$$C = \frac{X_0}{1 - \dfrac{m\Omega^2}{k}}$$

But

$$\omega = \sqrt{\frac{k}{m}}$$

$$\frac{k}{m} = \omega^2$$

$$\frac{m}{k} = \frac{1}{\omega^2}$$

so

$$C = \frac{X_0}{1 - \dfrac{m\Omega^2}{k}}$$

$$C = \frac{X_0}{1 - \dfrac{\Omega^2}{\omega^2}}$$

We can use this equation to determine the resulting amplitude of the system vibration, C (m).

$$C = \frac{X_0}{1 - \left(\dfrac{\Omega}{\omega}\right)^2} \qquad\qquad (8.5)$$

where X_0 = forcing displacement amplitude (m)

Ω = forcing frequency (rad s^{-1})

C = response amplitude (m)

ω = natural frequency of the system (rad s^{-1}).

$\dfrac{\Omega}{\omega}$ = frequency ratio $\dfrac{1}{1 - \left(\dfrac{\Omega}{\omega}\right)^2} = T$,   transmissibility function

Conventionally the transmissibility function always takes a positive value.

So the response amplitude of the system is related to the forcing displacement amplitude by:

$$C = TX_0 \qquad\qquad (8.6)$$

It is interesting to look at the transmissibility function more closely. Plotting the transmissibility function against the frequency ratio produces the following graph:

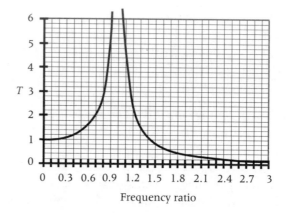

From the graph it can be seen that when the forcing frequency is much lower than the natural frequency, for values of frequency ratio up to about 0.3, then the transmissibility is around 1, meaning that the response amplitude will be approximately equal to the forcing displacement amplitude. For frequency ratio values above 1.5, that is when the forcing frequency is much higher than the natural frequency, the transmissibility is small and therefore the response amplitude will be considerably less than the forcing displacement amplitude. As the frequency ratio approaches 1, the transmissibility rapidly increases, with the result that the response amplitude becomes much larger than the forcing displacement amplitude. In fact at a frequency ratio of 1, the transmissibility becomes infinite. In this condition the forcing frequency is equal to the natural frequency and is referred to as resonance. For real systems, damping will moderate this somewhat but even then, the system displacement at resonance will be very large.

EXAMPLE 8.2

The spring/mass system described in Example 8.1 is subjected to a sinusoidal ground movement which has an amplitude of 0.15 m. Determine the response amplitude of the system if the forcing frequency is: (a) $2\,\mathrm{rad\,s^{-1}}$, (b) $15\,\mathrm{rad\,s^{-1}}$, (c) $7\,\mathrm{rad\,s^{-1}}$.

Solution

In each case calculate the frequency ratio $\dfrac{\Omega}{\omega}$ and hence the response amplitude.

In Example 8.1 we found the natural frequency of the system to be $7.071\,\text{rad s}^{-1}$.

$$C = \frac{X_0}{1 - \left(\dfrac{\Omega}{\omega}\right)^2}$$

(a) $\dfrac{\Omega}{\omega} = \dfrac{2}{7.071} = 0.283$ (b) $\dfrac{\Omega}{\omega} = \dfrac{15}{7.071} = 2.121$ (c) $\dfrac{\Omega}{\omega} = \dfrac{7}{7.071} = 0.999$

$C = \dfrac{0.15}{1 - 0.283^2}$ $C = \dfrac{0.15}{1 - 2.121^2}$ $C = \dfrac{0.15}{1 - 0.999^2}$

$C = 0.16\ \text{m}$ $C = 0.04\ \text{m}$ $C = 75.04\ \text{m}$

8.3 Forced vibration by applied force: one degree of freedom, undamped systems

Forced vibration can also be introduced to a system through the application of a sinusoidal force directly to the mass. The force has an amplitude, F_0 and frequency, Ω. Details of the arrangement are shown in Figure 8.4(a) and the free-body diagram of the mass is shown in Figure 8.4(b).

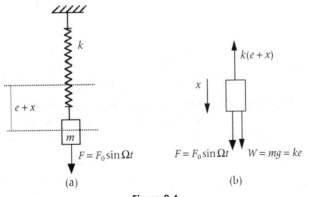

(a)   (b)

Figure 8.4

Applying Newton's second law to Figure 8.4(b):

$$\Sigma F_x = m\ddot{x}$$
$$ke - k(e + x) + F = m\ddot{x}$$
$$m\ddot{x} + kx = F$$
$$m\ddot{x} + kx = F_0 \sin \Omega t$$

If we compare this equation with the one developed for the ground motion model we find that the only difference is that kX_0 has been replaced by F_0. There is no need to rework the development of the response amplitude equation, we need only replace X_0 in the final result by F_0/k. The response amplitude equation for an applied sinusoidal force is given by:

$$C = \frac{\dfrac{F_0}{k}}{1 - \left(\dfrac{\Omega}{\omega}\right)^2}$$

(8.7)

where F_0 = driving force amplitude (N)
Ω = forcing frequency ($\mathrm{rad\,s^{-1}}$)
k = system stiffness ($\mathrm{N\,m^{-1}}$)
C = response amplitude (m)
ω = natural frequency of the system ($\mathrm{rad\,s^{-1}}$).

$\dfrac{\Omega}{\omega}$ = frequency ratio $\dfrac{1}{1 - \left(\dfrac{\Omega}{\omega}\right)^2} = M$,   magnification factor

As was the case with the transmissibility function, the magnification factor conventionally always takes a positive value. For undamped systems the transmissibility function and the magnification factor are identical. However this is not so for damped systems and hence the need for the two terms. Damped systems are beyond the scope of this text.

EXAMPLE 8.3

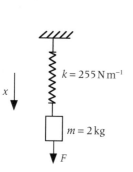

$k = 255\,\mathrm{N\,m^{-1}}$

x

$m = 2\,\mathrm{kg}$

F

A system modelled as a spring of stiffness $255\,\mathrm{N\,m^{-1}}$ and a single mass of $2\,\mathrm{kg}$ is subjected to a sinusoidal force, F, of amplitude $25\,\mathrm{N}$ and frequency $30\,\mathrm{Hz}$. Calculate the amplitude of the vibration of the system.

Solution

Determine the natural frequency of the system and then use the applied force response amplitude equation. Take care to ensure that the forcing frequency and the natural frequency are both expressed in the same units, either $\mathrm{rad\,s^{-1}}$ or Hz.

$$\omega = \sqrt{\frac{k}{m}}$$
$$\omega = \sqrt{\frac{255}{2}}$$
$$\omega = 11.292\ \mathrm{rad\,s^{-1}}$$
$$\Omega = 30\ \mathrm{Hz}$$
$$\Omega = (3 \times 2\pi)\ \mathrm{rad\,s^{-1}}$$
$$\Omega = 18.850\ \mathrm{rad\,s^{-1}}$$

$$C = \frac{\dfrac{F_0}{k}}{1 - \left(\dfrac{\Omega}{\omega}\right)^2}$$

$$C = \frac{\dfrac{25}{255}}{1 - \left(\dfrac{18.850}{11.292}\right)^2}$$

$$C = 0.0548 \text{ m}$$

The amplitude of the vibration resulting from the sinusoidal driving force is 55 mm.

8.4 Further discussion on examples

Although we have limited ourselves in this introduction to systems consisting of a single mass and single spring with no damping effects, these simple models do in fact yield reasonable results for a wide range of practical situations. More accurate models can be developed if the system is approximated by a series of interconnected springs and masses and if damping is taken into account, but such models are beyond the scope of this text. Example 8.1 shows how the natural frequency of vibration of a simple system can be calculated and how such a system would react to an initial displacement from its equilibrium position. Examples 8.2 and 8.3 investigate the resulting response amplitude of simple systems to harmonic forcing vibrations due to ground motion or a driving force.

8.5 Problems

8.1 Determine the natural frequency of vibration and period of the mass/spring system shown on the right.

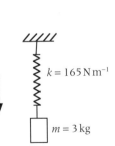

8.2 A mass/spring vibration system is to be designed to have a period of 3 s. If the system mass is 2 kg, determine the required stiffness for the spring.

8.3 If the period of the system described in **8.2** is to be doubled while the mass remains as 2 kg, determine the corresponding change in the required spring stiffness.

8.4 If the system shown in **8.1** is released from a point 120 mm below its steady-state equilibrium position with a downward speed of 2 m s^{-1}, sketch the resulting vibration and state the subsequent amplitude.

8.5 If the system shown in **8.1** is again released from a point 120 mm below its steady-state equilibrium position but this time with an upward speed of $2\,\text{m}\,\text{s}^{-1}$, sketch the resulting vibration and state the subsequent amplitude.

8.6 The system described in **8.1** is subjected to a sinusoidal ground motion, X, which has an amplitude of 0.1 m and frequency of $6\,\text{rad}\,\text{s}^{-1}$. Determine the response amplitude of the system.

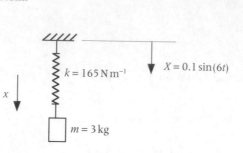

$k = 165\,\text{N}\,\text{m}^{-1}$

$X = 0.1\sin(6t)$

x

$m = 3\,\text{kg}$

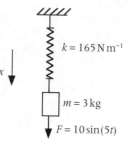

$k = 165\,\text{N}\,\text{m}^{-1}$

x

$m = 3\,\text{kg}$

$F = 10\sin(5t)$

8.7 The system described in **8.1** is now subjected to a sinusoidal driving force, F, of amplitude 10 N and frequency $5\,\text{rad}\,\text{s}^{-1}$ (as shown on the left). Determine the response amplitude of the system.

8.8 A mass/spring system is subjected to sinusoidal ground motions which can be described by the equation $X = 0.3\sin(\Omega t)$ as shown below.
(a) Calculate the resonant frequency of the system.
(b) Determine the response amplitude for a forcing frequency of $3\,\text{rad}\,\text{s}^{-1}$.
(c) Determine the response amplitude for a forcing frequency of $5\,\text{rad}\,\text{s}^{-1}$.

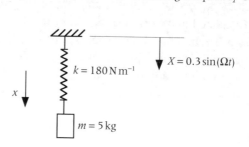

$k = 180\,\text{N}\,\text{m}^{-1}$

$X = 0.3\sin(\Omega t)$

x

$m = 5\,\text{kg}$

Answers

1 Initial concepts

1.1 Mass = 3298.67 kg, weight = 32.36 kN

1.2
(a) −63.1 kN m

(b) −200.26 kN m

(c) −138.065 kN m

(d) −10 kN m

(e) 116.14 kN m

(f) 70 kN m

1.3
(a) $\dfrac{dy}{dx} = 36x - 4$

(b) $\dfrac{dy}{dx} = 36x^2 - \dfrac{60}{x^4}$

(c) $\dfrac{dy}{dx} = 15(3x + 1)^4$

(d) $\dfrac{dy}{dx} = 12x(2x^2 + 5)^2$

(e) $\dfrac{dy}{dx} = -\dfrac{10}{(2x + 5)^6}$

(f) $\dfrac{dy}{dx} = \dfrac{6x + 1}{2\sqrt{3t^2 + t - 1}}$

(g) $\dfrac{dy}{dx} = \dfrac{1.8x}{(1 + 3x^2)^{0.7}}$

(h) $\dfrac{dy}{dx} = 10x^9(1 - 3x)(1 - 2x)^4$

1.4
(a) $\dfrac{dy}{dx} = 9x^2 + 4x - 3$

(b) $\dfrac{dy}{dx} = 1 - 2x$

(c) $\dfrac{dy}{dx} = 3x^2$

1.5 (a) $\dfrac{dy}{dx} = -\dfrac{2}{(x-1)^2}$

(b) $\dfrac{dy}{dx} = \dfrac{2(x^3 + 6x^2 + 12)}{(x+4)^2}$

(c) $\dfrac{dy}{dx} = \dfrac{3(9 - 4x^2)}{(x-1)^4}$

(d) $\dfrac{dy}{dx} = \dfrac{3x^2(x^2 + 2x + 5)}{(3x^3 - 15x)^2}$

1.6 (a) $\dfrac{dy}{dx} = 4\cos(x)$

(b) $\dfrac{dy}{dx} = 4\cos(4x)$

(c) $\dfrac{dy}{d\theta} = 1 + 2\sin(\theta)$

(d) $\dfrac{dy}{dx} = 3\cos(3x + 2)$

(e) $\dfrac{dy}{dx} = 6\cos(2x) - 4\sin(2 - x)$

(f) $\dfrac{dy}{dx} = 6\cos\left(3x - \dfrac{\pi}{2}\right)$

(g) $\dfrac{dy}{dt} = 2\pi\sin(1 - 2\pi t)$

(h) $\dfrac{dy}{dx} = 9(2x - 6)\sin^2(x^2 - 6x)\cos(x^2 - 6x)$

1.7 (a) $\dfrac{dy}{dx} = 3\theta\cos(\theta) + 3\sin(\theta)$

(b) $\dfrac{dy}{dx} = 2z\cos(z) - z^2\sin(z)$

(c) $\dfrac{dy}{dx} = 2(t^2 + 1)\cos(2t) + 2t\sin(2t)$

(d) $\dfrac{dy}{dx} = (x^2\cos(x))(3\cos(x) - 2x\sin(x))$

(e) $\dfrac{dy}{dx} = \dfrac{3(t\cos(t) - 4\sin(t))}{t^5}$

(f) $\dfrac{dy}{dx} = \dfrac{(1 + \theta)\cos(\theta) - \sin(\theta)}{(1 + \theta)^2}$

(g) $\dfrac{dy}{dx} = \dfrac{1}{(1 + \cos(x))}$

(h) $\dfrac{dy}{dx} = \dfrac{3\cos(3\theta)\cos(2\theta) + 2\sin(3\theta)\sin(2\theta)}{\cos^2(2\theta)}$

1.8 (a) $\dfrac{dy}{dx} = 3e^{3x}$

(b) $\dfrac{dy}{dx} = 2e^{2x+1}$

(c) $\dfrac{dy}{dx} = 2(3x+2)e^{3x^2+4x}$

(d) $\dfrac{dy}{dx} = e^x(1+x)$

(e) $\dfrac{dy}{dx} = \dfrac{xe^x}{(x+1)^2}$

(f) $\dfrac{dy}{dx} = e^{2\theta}(2\cos(\theta) - \sin(\theta))$

(g) $\dfrac{dy}{dx} - e^{2x}(3\cos(3x) + 2\sin(3x))$

(h) $\dfrac{dy}{dx} = \dfrac{(\sin(x) + \cos(x))}{e^x}$

(i) $\dfrac{dy}{dx} = e^{\sin(x)\cos(x)}\cos(2x)$

(j) $\dfrac{dy}{dx} = \dfrac{1}{x}$

(k) $\dfrac{dy}{dx} = \dfrac{1}{(1+x)}$

(l) $\dfrac{dy}{dx} = \dfrac{6}{(2+3x)}$

(m) $\dfrac{dy}{dx} = \dfrac{2}{x}$

(n) $\dfrac{dy}{dx} = \cot(\theta)$

(o) $\dfrac{dy}{dx} = 2\cot(\theta)$

(p) $\dfrac{dy}{dx} = \dfrac{1}{x(x+1)}$

(q) $\dfrac{dy}{dx} = \dfrac{6}{t}$

(r) $\dfrac{dy}{dx} = 2\sec(x)$

1.9 (a)

x	f(x)	f'(x)	f''(x)	f'''(x)
3	14			
3.1	17.252	34.08		
3.2	20.816	37.26	32.4	
3.3	24.704	40.56	33.6	12
3.4	28.928	43.98	34.8	12
3.5	33.5	47.52	36	12
3.6	38.432	51.18	37.2	12
3.7	43.736	54.96	38.4	12
3.8	49.424	58.86	39.6	
3.9	55.508	62.88		
4	62			

(b)

x	f(x)	f'(x)	f''(x)	f'''(x)
3	6.3504			
3.1	1.998	−46.696		
3.2	−2.9888	−52.9365	−61.0575	
3.3	−8.5893	−58.9075	−57.7225	39.85
3.4	−14.7703	−64.481	−53.0875	52.9125
3.5	−21.4855	−69.525	−47.14	65.9625
3.6	−28.6753	−73.909	−39.895	78.7625
3.7	−36.2673	−77.504	−31.3875	91.0625
3.8	−44.1761	−80.1865	−21.6825	
3.9	−52.3046	−81.8405		
4	−60.5442			

(c)

x	f(x)	f'(x)	f''(x)	f'''(x)
3	20.0855			
3.1	22.1979	22.235		
3.2	24.5325	24.5735	24.6125	
3.3	27.1126	27.1575	27.2025	27.275
3.4	29.964	30.014	30.0675	30.1125
3.5	33.1154	33.171	33.225	33.25
3.6	36.5982	36.659	36.7175	36.7875
3.7	40.4472	40.5145	40.5825	40.6625
3.8	44.7011	44.7755	44.85	
3.9	49.4023	49.4845		
4	54.598			

1.10 (a) $y = x^4 - x^3 + 6x + C$

(b) $y = x^3 + \dfrac{2x^3}{3} + C$

(c) $y = \dfrac{x^4}{4} + C$

(d) $y = -\dfrac{x^3}{5} + C$

(e) $y = x^2 + \dfrac{5}{x^2} - \dfrac{2}{x} + 2\ln|x| + C$

1.11   (a) $y = \dfrac{1}{2}\ln|x^2 + 6x| + C$

(b) $y = \dfrac{5e^{3x}}{3} + C$

(c) $y = \dfrac{4\cos(3x)}{3} + C$

1.12 $y = x\sin(x) + \cos(x) + C$

1.13   (a)   172

(b)   0

(c)   244.167

1.14   (a)   173.675 (Simpson's rule), 173.296 (Trapezoidal rule)

(b)   0 (Simpson's rule), 0 (Trapezoidal rule)

(c)   244.167 (Simpson's rule), 245 (Trapezoidal rule)

1.15

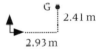

1.16 $I_{xx} = 0.01699\,\text{kg}\,\text{m}^2$ $I_{yy} = 0.1693\,\text{kg}\,\text{m}^2$ $I_{zz} = 0.00021\,\text{kg}\,\text{m}^2$
$k_{xx} = 0.130\,\text{m}$ $k_{yy} = 0.130\,\text{m}$ $k_{zz} = 0.14\,\text{m}$

1.17   (a)

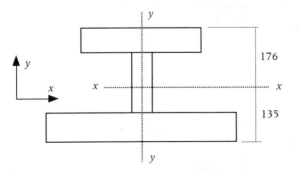

(b) $5.4558 \times 10^8\,\text{mm}^4$

(c) $2.3569 \times 10^8\,\text{mm}^4$

1.18   5123.9 kg

2 Kinematics

2.1 (a) 0, 9, 12, 27, 72, 165 m

(b) 18, 3, 6, 27, 66, 123 m s^{-1}

(c) 2 m s^{-1}, 1.333 s

(d) 66 m s^{-2}

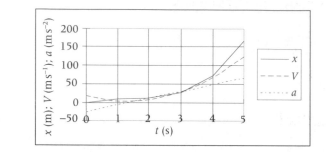

2.2 48 mins

2.3 3.84 m s^{-1}, 22.48 m s^{-1}

2.4 (a) 98.952 m

(b) 108.352 m

2.5 −0.83 m s^{-2}, 72.96 m

2.6 (a) 9.165 m s^{-1}

(b) 10.536 m s^{-1}

(c) 4.57 s

2.7 1.21 m

2.8 19.12 rad s^{-1}, 1.85 s

3 Newton's laws of motion: translational motion

3.1 290.363 N

3.2 458.301 N

3.3 6.313 m s^{-2} →

3.4 199.73 N

3.5 2.264 m s^{-2}, 60.37 N

3.6 5.244 m s^{-2}, 22.83 N

3.7 $P = 51.773$ N tie, $Q = 73.206$ N tie

3.8 $P = 57.735$ N tie, $Q = 115.47$ N tie

3.9 $P = 57.735$ N strut, $Q = 57.735$ N strut

3.10 $P = 108.195$ N tie, $Q = 124.840$ N strut

3.11 $F_{CD} = 3.750\,\text{kN}$ tie, $F_{BD} = 6.250\,\text{kN}$ strut, $F_{CB} = 15.373\,\text{kN}$ strut, $F_{AC} = 6.988\,\text{kN}$ strut, $F_{AB} = 3.125\,\text{kN}$ tie

4 Newton's laws of motion: rotational motion

4.1 $40.85\,\text{rad}\,\text{s}^{-2}$ clockwise, $1987\,\text{N}\ 75.4°\ \searrow$

4.2 $0.218\,\text{m}\,\text{s}^{-2}\ \leftarrow$, $0.581\,\text{rad}\,\text{s}^{-2}$ anticlockwise

4.3 $0.335\,\text{m}\,\text{s}^{-2}$, $37.888\,\text{N}$

4.4 $R_A = 22.542\,\text{kN}$, $R_B = 24.458\,\text{kN}$

4.5 $R_A = 40.6\,\text{kN}$, $R_B = 13.4\,\text{kN}$

4.6 $R_A = 46.833\,\text{kN}$, $R_B = 60.167\,\text{kN}$

4.7 $R_A = 30.706\,\text{kN}$, $R_B = 25.294\,\text{kN}$

4.8 $F_{CD} = 3.750\,\text{kN}$ tie, $F_{CB} = 15.373\,\text{kN}$ strut, $F_{AB} = 3.125\,\text{kN}$ tie

4.9 $F_{AC} = 15\,\text{kN}$ strut, $F_{AD} = 7.071\,\text{kN}$ strut, $F_{BD} = 20\,\text{kN}$ strut, $F_{AB} = 20\,\text{kN}$ tie

4.10 $F_{DE} = 21.2\,\text{kN}$ tie, $F_{DB} = 1.342\,\text{kN}$ tie, $F_{BC} = 8.1\,\text{kN}$ tie

5 Work–energy methods

5.1 $253.103\,\text{J}$

5.2 $365.351\,\text{kJ}$

5.3 $1.584\,\text{m}$

5.4 (a) $0.6\,\text{J}$

     (b) $1.8\,\text{J}$

5.5 $43\,671\,\text{kg}$

5.6 $\mu \geq 0.06$, $v = 3.263\,\text{m}\,\text{s}^{-1}$, $\omega = 21.756\,\text{rad}\,\text{s}^{-1}$

5.7 $1.379\,\text{m}\,\text{s}^{-1}\ \uparrow$

5.8

x (m)	Spring state (m)	ω (rad s^{-1}) clockwise
0	0.25 compressed	0
0.05	0.20 compressed	6.709
0.10	0.15 compressed	8.945
0.15	0.10 compressed	10.248
0.20	0.05 compressed	10.955
0.25	free length	11.181
0.30	0.25 stretched	10.955
0.35	0.20 stretched	10.248
0.40	0.15 stretched	8.945
0.45	0.10 stretched	6.709
0.50	0.05 stretched	0

6 Impulse–momentum methods

6.1 $10\,\text{Ns} \rightarrow$

6.2 $66.14\,\text{Ns} \rightarrow$

6.3 (a) $2.354\,\text{ms}^{-1} \uparrow$

(b) $5.351\,\text{ms}^{-1} \uparrow$

6.4 $6.2\,\text{s}$

6.5 $2.068\,\text{ms}^{-1} \rightarrow$

6.6 $100.584\,\text{N}$

6.7 $V_A = 0.8\,\text{ms}^{-1} \rightarrow$, $V_B = 2.6\,\text{ms}^{-1} \rightarrow$

6.8 $V_A = 2.386\,\text{ms}^{-1}\ \angle31.18°$, $V_B = 3.495\,\text{ms}^{-1}\ \angle13.26°$

7 Beam equations

7.1

x (m)	M_x formula	M_x (kN m)
0	$9.3x$	0
1		9.3
2		18.6
2	$9.3x - 6(x - 2)$	18.6
3		21.9
4		25.2
5		28.5
6		31.8
7		35.1
7	$9.3x - 6(x - 2) - 15(x - 7)$	35.1
8		23.4
9		11.7
10		0

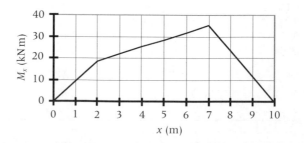

7.2

x (m)	M_x formula	M_x (kN m)
0	$19.8x$	0
1		19.8
2		39.6
2	$19.8x - 2(x - 2)^2$	39.6
3		57.4
3	$19.8x - 2(x - 2)^2 - 10(x - 3)$	57.4
4		61.2
5		61.0
6		56.8
7		48.6
8		36.4
9		20.2
10		0

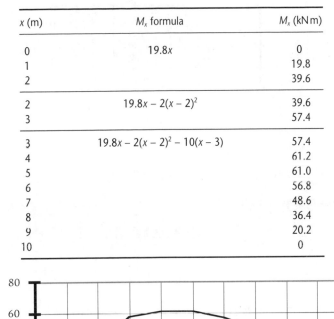

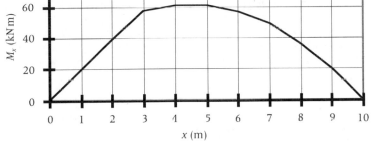

7.3

x (m)	M_x formula	M_x (kN m)
0	$-6x^2$	0
1		-6
2		-24
2	$-6x^2 - 10(x - 2)$	-24
3		-64
3	$-6x^2 - 10(x - 2) + 87.2(x - 3)$	-64
4		-28.8
5		-5.6
6		5.6
7		4.8
8		-8
8	$-6x^2 - 10(x - 2) + 87.2(x - 3) + 6(x - 8)^2 + 22.8(x - 8)$	-8
9		-4
10		0

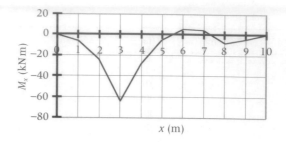

7.4 $x = 4.25\,\mathrm{m}$, $M_{\max} = 61.525\,\mathrm{kN\,m}$

7.5 $M_3 = -64\,\mathrm{kN\,m}$, $M_{6.433} = 6.727\,\mathrm{kN\,m}$, $M_8 = -8\,\mathrm{kN\,m}$

7.6 $x = 5.375\,\mathrm{m}$ and $7.492\,\mathrm{m}$

7.7 $0 \le x \le 2$ $\qquad\qquad\qquad\qquad$ $2 \le x \le 6$

$S_x = 16.333$ $\qquad\qquad\qquad\qquad$ $S_x = 1.333$

$M_x = 16.333x$ $\qquad\qquad\qquad$ $M_x = 16.333x - 15(x - 2)$

$\theta_x = \dfrac{1}{EI_A}\left(8.167x^2 - 118.214\right)$ $\qquad$ $\theta_x = \dfrac{1}{EI_A}\left(8.167x^2 - 7.5(x-2)^2 - 118.214\right)$

$Y_x = \dfrac{1}{EI_A}\left(2.722x^3 - 118.214x\right)$ $\qquad$ $Y_x = \dfrac{1}{EI_A}\left(2.722x^3 - 2.5(x-2)^3 - 118.214x\right)$

$6 \le x \le 9$

$S_x = -12.667$

$M_x = 16.333x - 15(x - 2) - 14(x - 6)$

$\theta_x = \dfrac{1}{EI_A}\left(8.167x^2 - 7.5(x-2)^2 - 7(x-6)^2 - 118.214\right)$

$Y_x = \dfrac{1}{EI_A}\left(2.722x^3 - 2.5(x-2)^3 - 2.333(x-6)^3 - 118.214x\right)$

x (m)	0	0.5	1	1.5	2	2.5	3	3.5	4	4.5
Y (mm)	0	−4.2	−8.3	−12.0	−15.3	−18.1	−20.3	−21.8	−22.8	−23.1

x (m)	5	5.5	6	6.5	7	7.5	8	8.5	9
Y (mm)	−22.8	−21.8	−20.1	−17.8	−14.9	−11.6	−8.9	−4.0	0

$Y_{\max} = Y_{4.492} = -23.1\,\mathrm{mm}$

7.8

x (m)	Y (mm)	θ (rad)	M (kNm)	S (kN)
0	0	−0.01122	0	35
1	−11.0	−0.01059	31.5	28
2	−20.8	−0.00889	56	21
3	−28.5	−0.00637	73.5	14
4	−33.4	−0.00331	84	7
5	−35.1	0	87.5	0
6	−33.4	0.00331	84	−7
7	−28.5	0.00637	73.5	−14
8	−20.8	0.00889	56	−21
9	−11.0	0.01059	31.5	−28
10	0	0.01122	0	−35

7.9

x (m)	Y (mm)	θ (rad)	M (kNm)	S (kN)
0	−61.2	0.06652	0	−10
0.5	−28.6	0.06266	−5.5	−12
1	0	0.05014	−12	−14 8.5
1.5	21.1	0.03562	−8.25	6.5
2	35.4	0.02580	−5.5	4.5
2.5	44.3	0.01923	−3.75	2.5
3	48.2	0.01446	−3	0.5
3.5	47.0	0.01005	−3.25	−1.5
4	39.8	0.00455	−4.5	−3.5
4.5	24.9	−0.00349	−6.75	−5.5
5	0	−0.01550	−10	−7.5
5.5	−36.8	−0.02590	−4.5	−9.5 10
6	−83.4	−0.02908	0	8

8 Vibration theory

8.1 $\omega = 7.416\,\mathrm{rad\,s^{-1}}$, $f = 1.180\,\mathrm{Hz}$, $\tau = 0.847\,\mathrm{s}$

8.2 $8.773\,\mathrm{N\,m^{-1}}$

8.3 1/4 of previous value

8.4 $x = 0.295\,\sin(7.416t + 0.419)$, amplitude $= 0.295\,\mathrm{m}$

8.5 $x = 0.295\,\sin(7.416t + 2.723)$, amplitude $= 0.295\,\mathrm{m}$

8.6 $0.290\,\mathrm{m}$

8.7 $0.111\,\mathrm{m}$

8.8 (a) $6\,\mathrm{rad\,s^{-1}}$
    (b) $0.4\,\mathrm{m}$
    (c) $0.982\,\mathrm{m}$

Index